CHAOTIC CHANGE

CHAOTIC CHANGE

EMBRACING CHAOS TO DRIVE INNOVATION AND GROWTH

NICHOLAS J. WEBB

☐ LeaderLogic®
COLEADERLOGIC.COM

Trademarks
Net Customer Experience™, Innovation Superstar®, Lucid Leadership®, RealRatings®, Happiness as a Strategy®, NickWebb®, and Net Customer Experience™. All other trademarks are the property of their respective owners.

Book illustrations by Stacy Ku.

Printed in the United States of America

ISBN 979-8-218-38681-8

First Printing

For general information, contact us at www.goleaderlogic.com.
To contact the author, visit www.nickwebb.com.

I would like to dedicate this book to my amazing family:
my wife, Michelle; and our children, Taylor, Madison, Chase, and Paige.

I would also like to further acknowledge my best friend for his
continuous support on all of my projects: Matti Palo, MD,
and all of my amazing clients who have trusted me over
the last thirty years to help them deliver astonishing strategic results.

Contents

Introduction

Change itself is nothing new. It's as old as the world. Since the dawn of time, people have both witnessed and championed the steady evolution of technology, medicine, culture, and science. For the most part, we've embraced change and made it work for us. We're smart, we humans! We're adaptable. And that's a good thing, because our eagerness to learn and make progress has been the key to our survival.

But as we plunge deeper into the twenty-first century, we sense that something is fundamentally different. The relentless stream of change we see and feel is more intense and less predictable than ever before. Our relationship to technology is evolving. For the first time in human history, we have built machines that think and act independently of human control. We have quantum computers that make decisions using a form of logic we don't understand. In the old days of the twentieth century, you would give a computer a set of instructions and a problem to solve, and it provided a predictable answer. Computers did what slide rules and abacuses did, only faster and with more power to handle complex calculations. Now we have machines that engage in self-directed learning, synthesize vastly more information than the human mind, and make their own choices among many variable, imperfect answers. It's both exhilarating and very scary. We see quantum technology and artificial intelligence (AI) as powerful genies let out of the bottle. There are even experts who assert that AI poses an "existential threat" to human life itself.

Can we control these powerful forces, or will they dominate and crush us?

We have entered the age of *chaotic change.* It's chaotic because the clear, analog-era rules of cause and effect have been erased. Well-defined conditions suddenly produce unforeseen results. Outcomes are less predictable. Instead of one clear answer to a problem, there may be many possible answers. Sometimes what we define as a problem may not be a problem at all!

Progress used to be like a river, flowing in one direction. You got in your boat and paddled with the current. In some places the flow was steady and calm, while in other places the water rushed through rocky rapids and you had to be highly skilled to stay afloat. But you always knew which direction you were going. Behind you was the past, and in front of you was the future.

In today's environment of chaotic change, you find yourself on a storm-tossed sea. The wind and waves come from all directions. Your compass is difficult to read. The stars are obscured by racing clouds. This is your new reality, and to survive you've got to let go of the old ways and embrace the new ways, and leverage them to your advantage. This is the essence of being a farsighted, bold leader—the willingness and ability to not only *accept* chaos but to *thrive* in it.

But there is another side to the story. Paradoxically, while chaos reigns in the world around us, many bedrock principles remain as immutable as ever.

Customers still want to buy what they need and be happy doing it. This has never changed.

Employees want fair pay and a positive, supportive workplace.

Investors want a good return on their money.

Leaders want to devote their time and effort to a project that has meaning.

The chaos, therefore, comes not so much in *what* you want to accomplish, but in *how* you can accomplish it. While the traditional goals remain, the strategies of yesterday have been swept away. The path to success has been obliterated, and it's time to blaze another.

Be a Chaotic Changemaker

Leaders who consistently excel in the marketplace and build scalable, predictable growth are not just *in tune* with change; they are obsessed with *anticipating it,* ensuring they're always several steps ahead of their competitors. Think of a grandmaster chess player strategizing a dozen moves in advance or top investors analyzing market patterns months ahead. Success hinges on foreseeing where the market, technology, and customer behavior will shift.

This foresight is the superpower of the *chaotic changemaker.*

The alternative is to be a *chaotic victim.* This is not a good choice for anyone.

When you're a chaotic changemaker, you seek every possible advantage to lead your organization to victory in an era of chaotic change. You've experienced significant shifts before, but something feels different now. You sense that the current wave of change is more erratic, asymmetrical, amorphous, and disparate. Yet, there's great news: Understanding the power of chaotic trends can equip you to build robust systems, not only to weather the storm but to harness these forces for unprecedented organizational success.

Chaotic Changemakers
Are Innovation Superstars

Chaotic changemakers grasp the true essence of innovation as we define it: "Innovation is the process of creating novel value that serves your organization and your customer." Central to this definition is the concept of "value." Historically, most organizations have aimed to achieve incremental or landmark value. However, chaotic changemakers forge paths toward breakthrough and disruptive innovations, leading the marketplace. While traditional organizations focus on maintaining the status quo, chaotic changemakers prioritize innovation, even if it means breaking the mold.

This philosophy may seem radical, but in an era of turbulent change, boldness emerges as the ultimate strategy for enterprises. It's often said, but it bears repeating: The greatest risk lies in avoiding risks. Striking the perfect balance between audacity and prudence is the key talent of exceptional leaders.

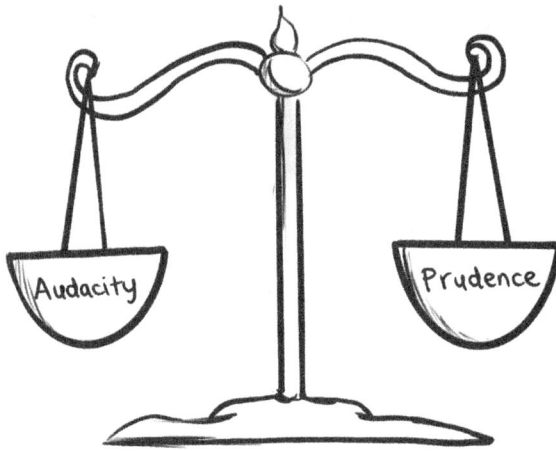

A key hallmark of today's chaotic changemakers is the understanding that fundamentally, most companies are *technology organizations*, regardless of what they sell or manufacture. This realization might sound drastic, but it's the reality of today's business landscape. Emerging and existing disruptive technology is crucial for delivering superior products, services, and human experiences. Even sellers of ubiquitous products such as concrete or paper know that the difference between profit and loss lies in their leveraging of technology to lower costs and boost efficiency. Today's leaders must understand this technological trajectory themselves rather than deferring to IT executives. *Chaotic Change* will reveal and discuss the critical importance of understanding the impact of emerging technologies on market value delivery.

Ensure Your Strategy Is Agile

As General Dwight D. Eisenhower famously said, "Plans are worthless, but planning is everything." In a world of chaotic change, nothing could be more true! Your enterprise operates on a strategy, but only agile, future-ready, innovative, and people-centric strategies will suffice. I'm not suggesting that your entire strategy is obsolete, but the methods to achieve predictable returns have evolved. The carefully constructed strategies of most organizations aren't designed for the velocity and magnitude of today's changes. Embracing and celebrating change, and aligning your pace and scale with the rhythm of change, are essential for meeting your goals.

You need accurate, timely information. How about surveys? The prevailing "survey industrial complex" of customer, employee, and market surveys has led many CEOs to believe their data is accurate and actionable. But this isn't always the case. True insight comes from understanding deeper needs, which will be discussed in this book. This book will introduce cutting-edge methods like hackathons, collaborative ideation, and enterprise innovation to provide you with the insights needed to maintain a competitive edge.

The Rise of Human Experience Innovation

Chaotic change is impacting your customers and employees, and the old approaches to them aren't working. Traditionally, customers and employees have been treated as if they were two different species of animal, with opposing desires and expectations. This is increasingly false. It's time to look beyond separate initiatives for customer and employee experiences to the comprehensive approach of Human Experience Innovation (HXI). Organizations often handle rising expectations incidentally and fail to integrate initiatives, leading to a disconnect between unhappy employees and the experiences of customers. HXI offers an integrated approach to foster a culture of enterprise happiness, driving sustainable growth and attracting essential talent.

Make Agility an Enterprise Core Competency

Many organizations are mired in slow, risk-averse processes that stifle their innovation pipeline. In a time of rapid, chaotic change, the greatest competitive edge is the ability to act quickly and boldly. This requires a high level of awareness, communication, teamwork, and the will to win. Organizations that have the *least* internal tension always have the *greatest* capacity for rapid and effective innovation.

Leaders and teams can't simply observe chaotic change through a narrow lens—be it new product development, workforce dynamics, leadership, or customer experience. Instead, you must consider all these factors holistically to implement the necessary organizational changes to profit from chaotic change. Moreover, cultivating a culture adept at managing change demands that it be embedded and entrenched systematically in every aspect of what you do. This encompasses how you devise future strategies and your approach to manufacturing, service delivery, messaging, and innovation. Essentially, every facet of your enterprise must be empowered with the systems of change mastery.

The Anatomy of Chaotic Change

Chaotic change is occurring systemically, yet oftentimes leaders assume that it's happening only within one area or department. To build an anticipatory enterprise that is "future ready," we need to look at change across as many systems as possible. In our research, we examine over 500 different trends, micro-trends, and nano-trends. These various trends are in a state of continuous flux, resulting in trends that change literally on a monthly basis. For the purpose of looking at the most conspicuous and pervasive trends, the following is a snapshot of some of the key trends that all leaders should be plugged into on a daily basis. This book will show you how to build the insights, systems, tools, and strategies to continuously adapt to the changes that are occurring across these key and atomic features.

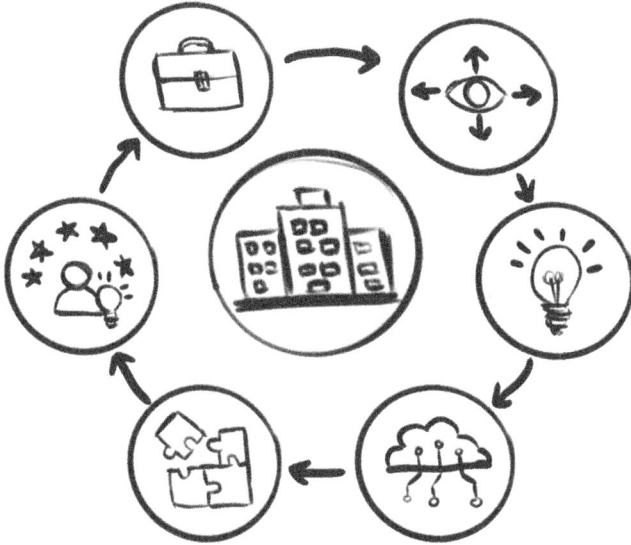

Customers. The evolving demands and behaviors of consumers in the market.

Employees. The changing dynamics and expectations of the workforce.

Strategy. The plans and actions companies undertake to navigate chaotic change.

Innovation. The new ideas and practices businesses implement to stay competitive.

Technology. The tools and systems that drive transformation in businesses.

Marketplace. The external environment where products and services are exchanged.

Enterprise. The comprehensive business structure and its adaptation to change.

Customers

Hyper-consumerism. Consumers' intensified demand for personalized and immediate products and services.

Experience economy. The trend of consumers valuing experiences over the acquisition of goods.

Subscription economy. The business model where customers pay a subscription price for ongoing access to products/services.

CX innovation. New approaches in customer experience to meet the high expectations of modern consumers.

Omni-channel engagement. A multichannel approach to sales that provides a seamless customer experience, whether the client is online or in a physical store.

Employees

Flex, remote work. The adoption of flexible work schedules and the ability for employees to work from outside the office.

Work-life integration. The blending of work and personal life for greater balance and well-being.

Upskilling and reskilling. The process of learning new skills or training in different fields to stay relevant in the job market.

Employee-based innovation. Encouraging and incorporating innovation originating from employees within the company.

Human Experience Innovation. The creation of new strategies to enhance the overall human experience within the workplace.

Strategy

Enterprise innovation. The implementation of innovative practices at the corporate level.

Human-centric strategies. Business strategies that prioritize the needs and experiences of people.

Anticipatory strategies. Strategic planning that anticipates future trends and prepares proactively.

Enabling technologies. Technologies that facilitate the execution of business strategies.

Future-ready planning. Preparing businesses for future challenges and opportunities.

Happiness as a strategy. Incorporating happiness and well-being as a key component of strategic planning.

Innovation

Beyond design thinking. Advancing past traditional design thinking to more innovative problem-solving methods.

Lean start-up methodology. A business approach that supports developing products that customers want quickly and efficiently.

Innovation as a strategic enabler. Using innovation as a primary tool to drive strategic success.

The collaborative enterprise. Encouraging collaboration within the enterprise to foster innovation.

Anticipatory innovation. Innovating with the anticipation of future needs and trends.

Innovation pipeline management. Overseeing the process through which new innovations are developed and brought to market.

Technology

Cloud and supercomputing. Utilizing cloud storage and powerful computing resources to process data.

Disruptive AI innovations. Artificial intelligence advancements that significantly alter industry landscapes.

Robotic process automation. The use of robots or AI to automate routine tasks.

Big data analytics. Analyzing large sets of data to uncover patterns and insights.

Digital and sensor ubiquity. The widespread use of digital interfaces and sensor technology in various applications.

Anticipatory technology models. Technology models that predict and respond to future operational needs.

Marketplace

Globalization. The increasing interconnection and interdependence of global markets and businesses.

Changing regulatory environment. The evolving landscape of laws and regulations that govern business operations.

Shifts in customer behavior. Changes in the way customers act, buy, and interact with brands.

Speed and boldness planning. Developing strategies that prioritize rapid action and courageous decision-making.

Experiential ecosystem planning. Designing business models that prioritize comprehensive, engaging customer experiences.

Beyond customer experience strategies. Approaches that surpass traditional customer experience tactics to offer deeper value.

Enterprise

The anticipatory enterprise. Organizations that proactively anticipate and respond to future challenges and trends.

Innovation as a core competency. Prioritizing innovation as a fundamental skill set within the company.

Beyond digital transformation. Pushing past traditional digital upgrades to more holistic and integrated transformations.

Strategic results through innovation. Achieving business goals primarily through innovative initiatives.

Human Experience Innovation. Innovating with a focus on enhancing the experiences of all people associated with the enterprise.

The new workforce. The evolving composition and expectations of the modern labor force, including attitudes, skills, and work habits.

The preceding list represents only a sample of the rapid changes that are occurring in the *marketplace*, within our own *culture*, and in our own sense of understanding and leadership clarity, and which require *lucid leadership*. In the pages ahead, the book will reveal how and why leadership itself has become the subject of rapid and chaotic change. This requires that organizational leaders become lucid leaders who lean into the blur of chaotic change to understand its DNA so they can lead it to enterprise success.

The Journey Through Chaotic Change

This book is divided into three parts.

Part I dives into the importance of effective leadership. It's paramount. No matter how many resources you have or the strength of your market position, if your organization lacks clear, decisive, ethical, growth-oriented leadership—what we call *lucid leadership*—then you'll have big problems. But fear not! *Chaotic Change* is streamlined and packed with practical steps you can take to strengthen your leadership and your organization. The first chapter presents a deep review of the importance of lucid leadership before presenting the Value Leadership Model, a step-by-step blueprint for leadership success. We'll then talk about the roles of directors and investors—they can make or break your company!—before tackling the powerful Make Innovation REAL system, which shows you how to make innovation the cornerstone of your chaotic change agenda.

No one can afford to ignore AI: It's here, it's becoming more ubiquitous, and every leader needs to know how to harness its power and control its liabilities. Part I concludes with what many may think is a statement of business heresy: You cannot focus only on what you think your customers want!

Part II focuses on the primary task of the leader, which is to promote innovation in all its forms. We define it as *the creation of new value that serves your organization's mission and customer*, and it must be woven into the fabric of the organization and sustained over time.

You begin by creating a *culture* of innovation, of which an important part is embracing happiness as a strategy, because unhappy employees don't innovate, they just take their paychecks and go home. This leads us to employee experience design and the innovation operating system, which show you how to make innovation a normal part of your daily operating procedures. You'll learn the three simple steps to driving innovation, the top ten innovation killers (watch out for them!), and finally the incredible value of business systems, with one important caveat: While you must have systems, they *must be agile and flexible*. While internal chaos may be bad, ossified, rigid business systems are just as deadly.

Part III reveals the importance of your human stakeholders—your people. The book effectively tears down the traditional barrier between employees and customers, and shows you that they're in fact two sides of the same coin! In many ways, your employees are your customers, while your customers share many of the same characteristics as your employees. This is Human Experience Innovation (HXI). It's your job to treat every customer and employee as a unique human being. While that lofty goal may be still in the future, you can break down foolish superficial stereotypes and see customers and employees according to what motivates them to either reject or support your mission.

Chaotic change is upon us! We can either greet this news with open arms or try to avoid it. We can either use it to fuel innovation or point to it as an excuse to retreat. The upcoming chapters will reveal how you and your organization can use chaotic change to your advantage and build the systemic aspects of enterprise development. This will ensure that you are equipped with the latest systems, tools, processes, and philosophies needed to excel amid chaotic change—which, as we'll discover, is just another name for opportunity.

Ready? Let's get started!

PART I

Leadership

Success in our time of chaotic change begins with leadership. Without effective leadership, no organization can hope to survive, much less thrive. Here are the key concepts and lessons that will help you strengthen your own leadership capabilities and instill the power of leadership in every corner of your organization.

The first and most important duty of a leader in a time of chaotic change is to create and sustain a record of profitability. Keep in mind that even if you're operating a not-for-profit or governmental agency, *you still need to be able to deliver measurable real value.* The foundation of profitability, and organizational value creation, is, after all, *innovation*—so let's begin our journey there.

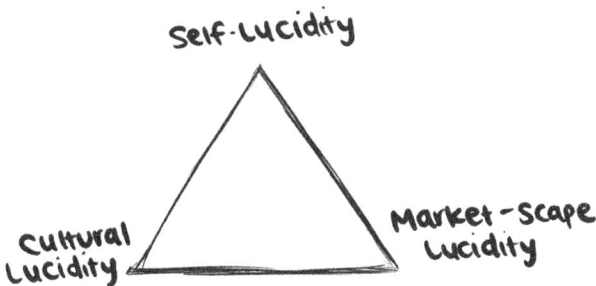

Self-Lucidity

Cultural
Lucidity

Market-Scape
Lucidity

Market-Scape Lucidity

The Trinity of Clarity

In our time of chaotic change, the best leaders leverage the benefits of clarity and lucidity. There is a trinity that makes up the three success drivers of leadership.

Self-Lucidity

The demands on leaders today are greater than ever. The skills that are required to operate in a time of hyper-complexity have rapidly evolved. Lucid leaders accept the challenge of learning, evolving, and ultimately growing. To be great as a leader, you must have a tremendous sense of self-awareness to understand where your competency gaps are so you can rapidly fill them to achieve a state of clarity.

Market-Scape Lucidity

Market-scape lucidity is the ability for a leader to go beyond just understanding the marketplace, in order to truly understand the entire landscape of the external market and all of the complex variables that will impact the organization. This includes understanding the trajectory of emerging technologies and how they will impact the enterprise. It also goes beyond outdated surveys and canned market studies. In order to achieve true market-scape lucidity, a chaotic leader must be obsessed with understanding all of the variables that determine the future of their enterprise. Lucid leaders are focused on understanding their customer, the impact of emerging technologies, and their own enterprise culture. The trinity of clarity is the ecosystem that great leaders live and thrive in.

Cultural Lucidity

And a time of rapid change, your success will be completely dependent upon your ability to attract, retain and inspire mission critical employees. Simply stated, this requires that you create a culture of enterprise, happiness, collaboration, and innovation. Creating a great culture begins with having an honest and candid understanding of what your culture is and what it is not. Successful leaders know their culture, and they are constantly moving it to a state of happiness and creativity.

It's important to remember that getting to a state of lucidity does require a genuine commitment to exploring the uncomfortable and the unfamiliar. This is the hallmark of the best innovators and the best leaders.

CHAPTER 1

Leverage Chaotic Change or Be Crushed by It

When you think about it, the world of business is a lot like pro football.

Consider the NFL. It comprises thirty-two teams. In the NFL marketplace, the teams compete for the cash generated from selling tickets and broadcast rights, and also for the prestige of winning the Super Bowl. It's really just like any other business environment. For example, in the automobile market, the big auto companies compete with each other for dominance. So do consumer electronics companies and fast-food companies in their respective markets. Each market is like a league, and in each league there can be only one champion.

Many teams, one champion.

Okay—so what does it take to be the champion?

Great leaders know it's all about *leveraging chaotic change*. Either you leverage it or you'll be crushed by it. There's no other choice.

Let's go back to football to see why.

Your company is like the offensive team. Your CEO is the quarterback. Your market rivals are the defensive team. During the game, every time the center snaps the ball, the quarterback faces the forces of chaotic change. Across the line of scrimmage, the defensive players—your business competitors—are determined to figure out which play is coming and how to smash it.

It's a constant battle between knowing what worked yesterday (i.e., respecting the tried and true) and embracing the necessity to adapt, improvise, and reinvent (i.e., leverage disruption).

If you ask an NFL head coach if he needs to be the master of chaotic change, he'll probably look at you like you've just stepped off a spaceship from Mars. Of course, NFL teams must leverage chaotic change! They do it every time they take the field. To win games in the NFL, you must constantly innovate. You must constantly read the competition. You must thrive in a world of chaotic change. When an injury sidelines your top receiver, you must adapt. When the opposing team tries a tricky new play, you need to figure out how to defeat it.

Every company has a strategic plan that guides its actions. The strategic plan of a pro football team is contained in its playbook. The typical pro football team has up to 1,000 plays in its playbook, and new ones are constantly being created. That's a lot of plays! Why so many? Because in order to score goals, the team needs deep organizational agility. They need to *act as disruptors* and turn the tables on the defense. The quarterback and his offensive team need to go beyond *enduring* chaotic change; they need to proactively *leverage* it to their advantage.

They must be unpredictable. Nimble. Agile. Always looking for the smallest advantage. And they need to do this while always remembering their core strengths.

Do CEOs Know What NFL Coaches Know?

Many CEOs love watching football and they admire winning coaches, while at the same time they can't see how the business of winning football games by *leveraging chaotic change* is just like increasing market share or boosting sales by always staying one step ahead of the competition.

True, there are many CEOs who recognize and try to respond to the steady pace of chaotic change in the marketplace. At least they're not hiding their heads in the sand! But they don't see that the *rate of change* is speeding up. Ten years ago, you could update your playbook once every season. That was enough to stay one step ahead of the competition. Today, chaotic change has accelerated and has become more powerful. You need to update your playbook as often as necessary. Every quarter, every month, every week—whatever it takes. You need to make chaotic change work *for* you, not *against* you.

The Five Building Blocks of Chaotic Change

Let's begin with a definition and description of chaotic change in the business marketplace. It comprises five key building blocks:

1. **Increased customer power.** Great leaders know their customers have more power than ever. People who buy your products have more choices, and they're more involved in the brands and suppliers they love (and hate!). They not only "vote" with their cash, but they can make their voices heard through social media and on platforms such as Yelp and Angie's List. They want and expect 100 percent on-time delivery in full and no-nonsense returns. No company can survive without managing the disruptive power of the customer—which means *exceeding* expectations, first time, every time.

2. **Revolutionary connection architecture.** Digital media has disrupted the way we do business. Internal tools now include intranets and real-time dashboards that can deliver up-to-the-minute company performance metrics and reveal problems demanding action. We now shop online, putting enormous pressure on brick-and-mortar retailers to innovate and redefine the customer experience. In the supply chain, the Internet of

Things (IoT) means that every step is tracked and verified—
there's no place for sloppy work to hide!

3. **Rapid innovation.** The pace of chaotic change is increasing.
Product life cycles are shrinking. Analysts say that across a range
of industries, 50 percent of annual company revenues are derived
from new products launched within the past three years. This
means that long-term product "cash cows," which may have been
in a company's portfolio for many years, are becoming a thing
of the past. And if a business is slow to introduce a new product
to market, it risks launching something that has already been
superseded by competitors.

4. **Evolving economic models.** Great leaders know the
days of the old top-down, hierarchical company are fading.
Today's digital world of work has shaken the foundation of
organizational structure, shifting from the traditional functional
hierarchy to a network of teams. As teams operate and customers
interact with the company, they must share information about
disruptions, what's working, what isn't working, and what
problems they need to address.

5. **Emerging employee personas.** Too many employers treat
their team members as if they live in a single demographic
bucket. In fact, leaders are increasingly recognizing a range of
what we call *employee personas*, based on what employees love
and hate. It's not about their age, income, position, ethnicity,
or any other traditional demographic factor. Great leaders know
how to identify their team members across a range of personas,
and by doing so they can engineer beautiful experiences for them
across a wide range of *team touchpoints*.

When a leader can be extremely granular about each of their team
members and understand them based on what they hate and love, they do
a far better job of engaging them, connecting with them, and ultimately
creating a work environment that delivers the best human experience for
them while delivering the best productivity and results for the enterprise.

Employees who feel like they're being treated with indifference become disengaged, less productive, and more likely to look for the exit.

Whew! How Can Anyone Keep Pace?

You may ask, how can any one person keep track of the relentless pace of chaotic change? Aren't there too many moving parts?

Relax. It's easy to assume that all of the changes happening across the changing customer, emerging technologies, and economic shifts are far too complicated to ever understand. The good news is that assumption is wrong. Great leaders develop a core competency around future trends. They don't need to understand the endless minutia of every little emerging technology, consumer behavior, or economic shift. If we can focus on these building blocks at a high level, we can begin to understand how these changes impact the way in which we deliver human experiences, technologies, and services, and ultimately how we serve to achieve sustainable results on strategy.

The "Baby-Step" Approach Doesn't Work

Great organizations sometimes attempt to leverage chaotic change by taking baby steps at disruptive initiatives that include future-casting activities, innovation customer experience (CX), workforce engagement initiatives—and the list goes on and on. The leaders therefore assume that by taking a cursory approach to leveraging chaotic change they'll be in pretty good shape. The problem is that the rate and depth of change requires a holistic and comprehensive approach toward gaining better customer and market insights, while doing a far better job of maximizing the efficiencies and effectiveness of our teams and leaders. In other words, today's disruption is changing everything, and how leaders lead is no exception.

"Wishful" and "Thinking" Don't Go Together

When the NFL quarterback gets the ball and steps back to survey the field, he's got about three seconds to assess his receivers and decide to whom he can throw a pass, or if it's better to try to pick up a few yards by running the ball himself.

His assessment of the field must be based on *stark reality*.

He cannot "wish" or "hope" or "assume" a receiver will be open when they're not.

He cannot ignore the three-hundred-pound linebacker headed toward him, intending to flatten him.

He can't hope he had the injured player who's sitting on the bench.

He must see the playing field as it really is *at that moment* and respond accordingly.

In business, great leaders know the first step toward leveraging chaotic change is to recognize the world as it really is and not as they'd like it to be.

They know that wishful thinking will put them on the road to disaster.

Organizations achieve success only *after* we've been honest and truthful about what is required to drive enterprise excellence. As a philosophical starting point, merely chasing "success" is misguided. If we aim only for success, we'll never get it. Success is the end result of staying focused on the truth.

This applies to how we manage our teams. We need to be honest about what we're asking them to do and then we need to address the realities of the challenges head-on. Chaotic change requires that we recognize what's really happening and respond decisively, without thinking, "We can't do that!"

There's nothing wrong with success—in fact, that's what this book is about! But great leaders know success is the result of the heavy lifting of leveraging chaotic change so they can ultimately win—and win *big*.

Don't Be a Dog Sitting on a Nail

There is a great story told by motivational speaker Les Brown about a traveler who was walking down the street and noticed a man and a dog sitting on a front porch. The dog was whimpering. The traveler approached the man on the porch and asked him why his dog was whimpering.

The man on the porch replied, "He's sitting on a nail."

The traveler then asked, "Why doesn't he just get up and move?"

The man replied, "I guess it isn't painful enough yet."

Many failing leaders are like the dog sitting on the nail. They're aware of the pain, but they'll wait until the organization and their career have suffered meaningful damage before they get up and move. Too many leaders live inside echo chambers surrounded by people who create safe spaces away from the outside world, and this makes the pain of chaotic change, at least in the short term, feel less painful.

Great leaders know when to get up and move! Not only does leveraging chaotic change remove pain points, it will also take you to a new and beautiful place, because chaotic change brings lot of good stuff that you can use to create superstar organizations and teams.

The Deception Chamber

Too many leaders are committed to *ignoring* chaotic change. They have created "deception chambers." These are the insulating circles of influence we build around us so we can feel more comfortable about the speed and depth of change.

Some CEOs will only hire team members who come from the same industry and who hold similar points of view. These CEOs have institutionalized the process of finding only those points of view and expertise that validate their self-constructed deception chamber.

It turns out, however, that if you want to lead a superstar organization, if you want to be relevant in a time of massive disruption, not only must you avoid echo chamber behavior, you need to seek out team members with an orthogonal point of view. Great leaders have developed teams made up of a tapestry of persona types, which help them develop the kinds of insights and collaborations that make for superstar organizations.

Don't be like the losing coach who thinks last year's playbook will work this year. Don't seek out opinions that you know reflect yesterday's thinking. Challenge yourself and those around you, and always demand to hear the truth. If you value truth over success, you'll be amazed at how easily true success will come!

The Three Strata of Self-Deception

The term "strata" refers to a range of layers, and in the area of self-deception there are three key layers of self-deception that failing leaders embrace. Understanding these three layers is important, because without understanding the very medium that makes up chaotic change, you'll never be able to protect yourself from it and leverage it for your benefit.

Here's a brief description of the three strata of self-deception.

1. The Echo Chamber Stratum

In this stratum, weak leaders make a formal commitment to eliminating any outside information or points of view that may be contrary to their personal prejudicial beliefs about change. They solicit the advice of internal team members and external partners and advisors who share their exact same frame of reference. This is very commonplace, and it's extremely dangerous.

Needless to say, ignoring or rejecting input that challenges your preconceptions is not a viable strategy. This is what the leaders of General Motors did for twenty years, right up until the company declared bankruptcy in June 2009. This was six months after the CEOs of Detroit's Big Three automakers—Rick Wagoner of GM, Alan Mulally of Ford, and Robert Nardelli of Chrysler—flew in their luxurious corporate jets to Washington, DC, to ask for billions in taxpayer bailout cash. In March 2009, Wagoner was fired, and his private jet put into a hanger.

Even for good leaders, the echo chamber problem can be unintentional and very difficult to solve.

2. The Data Stratum

The next problem we have—and it's a whopper—is the data stratum. This is where we begin the process of compiling conformational-based data. This is where organizations go out and specifically seek out data that *confirms their decision* to move forward on a safe and comfortable new product, technology, or even strategic direction. In fact, there is a massive multimillion-dollar enterprise that's in the business of getting data to make leaders and organizations feel better about the way in which they deliver enterprise value. As British Prime Minister Benjamin Disraeli put it: "There are three kinds of lies: lies, damned lies, and statistics." The information that you get related to the way in which you drive your organization and lead your teams is based plain and simply on *seeking the truth*.

3. The Self-Awareness Stratum

Living consciously and having accurate self-awareness are key principles driving superstar leaders. The flip side is the *self-awareness stratum*.

Deception is real and has a massive impact on virtually every decision you make on each and every day as a leader. This particular self-deception—that we know everything and that the massive changes currently happening are a short-term anomaly—is lethal to your career and your organization. Leaders begin from the level of self-awareness, and that means that every day they're conscious about how open they are to other people's ideas and the realities of the markets they serve.

You have to be careful not to be an "idea bully." Let's face it: You're a leader because you're smart and dynamic, and you strive to succeed. These are all great attributes but if left unchecked, anyone can become an authoritarian idea bully, and that constricts your abilities to get the fresh, objective insights you need to drive superstar organizations.

Take Action!

✗ Identify and evaluate the sources of *external* disruption in your market—the emerging forces that threaten your organization's viability. What might threaten your supply chain? What new competitors are rising? Are your customers coming back again and again, or are they defecting to competitors?

✗ Identify and evaluate the sources of *internal* disruption that are brewing right now within your organization.

✗ Are your employees fully engaged? Is your new product development on track? Are you personally open and receptive to new ideas?

✗ Make sure you're getting real information, not suppositions or platitudes. Do you seek critical viewpoints or subtly discourage them? Do you welcome chaotic change as a catalyst for improvement, or does it irritate you? Do you promote data that represents stark reality or data that makes you look good to your stakeholders?

CHAPTER 2

The Value Leadership Model

Many of the leadership models propagated today are incomplete and in some cases misleading. They're *fractional* and rooted in the past, which was a simpler time.

Today we're experiencing chaotic change and accelerating disruption. The world—and business—is growing more complex. We're putting behind us the days when one super-specialist could drive a company's success. Today's leaders must have both depth and breadth. They must be able to evaluate and make decisions across a broad range of business activities, including research and development, innovation, customer experience, marketing, finance, operations, human resources, strategic planning, and more.

And most of all—and this is critical—today's winning leader needs to forge *deep and productive relationships with his or her stakeholders.* No matter what industry you're in, you're first and foremost in the *people business.* There is no other way, and no shortcut. Either you do this or you lose touch and lose control—and quickly fall behind.

The Four Behaviors

To stay one step ahead, there are four behaviors you need to understand and apply every day. The four behaviors are:

1. **Inspire** your employees and stakeholders to excel.
2. **Connect** with them on a substantive, daily basis.
3. **Adapt** to rapid changes.
4. **Respect** your customers and stakeholders—and earn theirs in return.

Each behavior is supported by three action attributes. Together they make the Value Leadership Model.

Inspire

To stay ahead of chaotic change, it's imperative that as a leader you give your employees and stakeholders a reason to get up in the morning, come to work, and dedicate themselves to quality, innovation, and total customer satisfaction.

It's not the same thing as paying them high salaries or offering generous benefits. People can be very well paid and the same time be totally uninspired.

Inspiration is a form of *positive energy*. Winning leaders do not drag team members by the leash, nor do they threaten them into compliance. With crude tactics, no leader can inspire anyone in a meaningful and sustainable way.

Winning leaders create a powerful beacon that guides team members and draws them closer. Inspiration can be stronger than any other force. I'll bet you can think of a dozen global companies—Apple and Amazon are just two—that were started not with big paychecks or generous perks but only with a spirit of inspiration that drove the founders and their teams forward. They worked in garages or college dorm rooms, driven by inspiration, and hoping for—but not expecting—financial success. Most entrepreneurs do not achieve financial success, but they still reach for the stars, following their inspiration to innovate.

The ability to inspire is just one of the four behaviors necessary for winning leadership. But how do you do it? You can do it by understanding and mastering three attributes: vision, value, and viability.

1. Vision

Too many executives are not able to present a picture of how to recognize success. How can you endeavor a strategic initiative and drive strategic results when you haven't asked yourself the basic question, what does success look like?

To inspire others, you must first have a *vision* of where you want your people and the organization to be in the future. Your team members need to be able to visualize what success looks like and how their work will impact their community. The visualization must illuminate some positive change to people's lives.

If you have no vision for yourself or your organization, stop right now and start thinking about it.

In defining vision, winning leaders don't just about the "why"; it's much more. You have to fast-forward through the why to clearly visualize this in your mind: "If all goes as planned, what will our success look like?" Through visualizing the endgame, you significantly increase your ability to communicate a clear vision to your team and a path forward.

Before we go any further, we need to discuss "vision" versus "mission." Some people consider the terms to be interchangeable, but there's one important difference. While a mission statement describes what a company wants to do *now*, a vision statement outlines what a company *wants to be in the future*. A vision statement defines the core ideals that give a business long-term shape and direction. It's inspirational and aspirational, and creates a mental image of the future state that the organization wishes to achieve.

Some vision statements are very simple, like IKEA:
"To create a better everyday life for the many people."

Many companies have both a mission statement and a vision statement, like Airbnb:
Mission: "Belong anywhere."
Vision: "Tapping into the universal human yearning to belong—the desire to feel welcomed, respected, and appreciated for who you are, no matter where you might be."

And Tesla:
Mission: "To accelerate the world's transition to sustainable energy."
Vision: "To create the most compelling car company of the twenty-first century by driving the world's transition to electric vehicles."

2. Value

To inspire your team and stay ahead, you need to have more than just a vision of what success looks like. You need to also verify that if your vision were realized, it would actually be valuable to society.

It's incredible how many leaders obsess over missions that simply don't matter. They don't matter, because they're not valuable. They're not valuable to your team, to you, and most important, to your organization. Your team will only follow your vision if the vision makes sense and delivers real value.

For example, saying "It's our vision to make high profits" won't inspire anyone, because it's self-centered, meaning it says nothing about how your work impacts your community.

In contrast, saying, "It's our vision that everyone in America should live longer by eating healthy organic food" puts the spotlight on the effect your work will have and the value it will bring to your customers. Remember that today's employees are looking for more than a job; they are looking to participate in missions that matter.

Another common problem with a vision is that although it may have value, it may not be the number one priority, and chances are your team knows that. For example, saying "It's the vision of our fast-food company to sell burgers at the lowest price possible" may be one part of your business strategy (i.e., you compete on price), but it can't be your all-encompassing vision because it's a fragment. Anyone could sell the lowest-priced burger if it were tiny and made of crummy meat. By itself, it's a meaningless vision.

By comparison, in the low-price fast-food segment the mission statement of In-N-Out Burgers, Inc., is, "Give customers the freshest, highest quality foods you can buy and provide them with friendly service in a sparkling clean environment."

It focuses on the overall customer experience, which is something that every employee can understand.

Winning leaders have a well-defined vision of what success looks like, and they confirm and communicate the real value of a mission that matters.

3. Attainability

A multibillion-dollar organization was struggling to recover from a serious management misstep. During an executive strategy session, top leaders had decided the company needed to increase revenue by 15 percent in the coming year. That was all: "Just get revenues up!"

To accomplish this goal, they brought in their internal branding team and other key executives to build out what they called their "Fast 15 Strategy." To support the strategy they created logos, slogans, communication memos, launch meetings, and other messaging materials.

There was a problem with this frenzy of activity: There was no logical way the organization could achieve this rapid growth. There was no strategy other than to demand their employees magically make it happen. In fact, their ambitious revenue strategy was proposed in the midst of a declining market segment, and most market indicators indicated that the company would actually see reduced revenues in the coming year.

Of course, the initiative failed. While the resulting revenue shortfall was bad, more important, the top leaders had done something deadly to their leadership credibility—they put together a strategic initiative around growth that was not attainable, and *their teams knew it. To inspire others, you must present attainable goals.* Protect your leadership brand at all costs by pushing out only directives and initiatives that are truly valid.

Connect

Many economists say that because of the internet and global digital communications, we live in a "connected economy." Although it's true that we have more fibers of digital connectivity than ever before in human history, sociologists argue the opposite—that in fact we're a *disconnected* society composed of people who have fewer and fewer human interactions.

So which is it: Are we a connected people or a disconnected people?

Well, have you ever gone to a restaurant and seen a family sitting at the table, with each member staring down at their personal digital device? They can send text messages, emojis, photos, and even money to each other, all without talking or even looking at each other.

Are they really connected?

And as a leader who wants to stay ahead, are you really connected to your team, your stakeholders, your customers, your vendors? Or are you living in a digital simulation of reality?

I believe we're a *selectively* connected society that is largely driven by the way we architect engagement. We can choose to engage virtually through digital media or the old-fashioned way of face time. In this case, it's a good idea to strike a balance between the old and the new.

Being connected means having a *personal relationship* with your employees and other stakeholders. No, you don't have to know every detail of their private lives, but you should have a mental picture of as many people as possible in your organization. You need to see them as individuals, not as cogs in the machine.

The Connection Quiz

To help you strengthen your personal connections, here's a little quiz that you should give yourself on a regular basis. The quiz takes the form of ten simple questions.

For every question, *email does not count.* Emailing a subordinate counts as *zero* on the connection scale!

You should self-administer Part 1 of the quiz at the *end of the workday*:

1. How many subordinates did I see face-to-face today?
2. How many subordinates did I speak with one-to-one today?
3. How much time did I spend out of my office, visiting different areas of the organization and meeting with the people who staff them?
4. How many subordinates did I meet for the first time today?

Part 2 of the quiz you should self-administer *every week*:

1. How many of my subordinates can I name by sight?
2. With how many of my subordinates have I exchanged a personal greeting?
3. How much time, in hours or minutes, did I spend directly interacting with subordinates?
4. How many suggestions submitted by subordinates have I seen and reviewed this week?

Over time, you should increase the numbers of the responses to every question. Every time you increase your connection to your people—which boosts employee engagement, which is another emerging key metric—you make your organization stronger and more resilient.

Three Key Attributes

There are three key attributes in the second Value Leadership Model behavior that winning leaders know and practice. They are access, action, and authenticity:

1. Access

When it comes to team engagement, providing access goes far beyond the so-called open-door policy. Having an "open door" means your team members can enter your office when they want to, to ask a question or deliver a report. This is nice, and it's better than having a closed door, but it means you are *passively* waiting for them to interact with you.

Winning leaders *proactively* reach out to engage team members in a way that is relevant to their individual archetypes, goals, and responsibilities. It means leaving your office and intruding into their space (nicely, of course!) to ask them how they're doing and—more important—what you can do to help them get the job done.

Remember, while your employees work for you, in a very real sense you are their servant. Your job is to help them do theirs in the best and most productive way possible.

It means recognizing and adapting your interaction to fit the *archetype type* of the stakeholder. Every human being is unique. Some are extroverted, others introverted. Some are impulsive, others deliberative. A winning leader knows how to interact with each type so that a deeper, more lasting connection is made.

Remember, you can't communicate vision, value, and viability if you're not personally outreaching to the teams you serve. And this does not mean sending tweets. It means putting in real face time. In the old days they called it "management by wandering around," and today it's more necessary than ever.

For example, Don Fox, chairman and former CEO of Jacksonville, Florida-based Firehouse of America, parent company of Firehouse Subs, makes visiting his franchise restaurants high on his annual priority list, which he compiles at the beginning of each year. Nurturing relationships within the franchise community is a big part of his job. He told Julie Knudson from *QSR* magazine that the primary benefit of a hands-on approach is that "it keeps you very close to the business, how the business is being operated, and where the dollars are flowing." If too much of your time revolves around headquarters, Fox told Knudson, you're probably missing out on critical information, including how customers interact with your brand.

2. Action

Team engagement is a dynamic and active process. To engage and connect is to lead. The problem is that most leaders are busy, and the idea of getting up out of the corner office and actively setting up processes of systematically engaging team members is tiresome. In fact, some leaders may consider this to be a waste of time. That is unfortunate, because there is no better way to drive enterprise success than to increase the efficiencies and engagement of your team members. Action is the heavy lifting of leadership, and it requires that you build processes and routines around the active connection of your team.

Winning leaders transform plans into reality. *Too often, CEOs propose impressive strategies and plans that sound good on paper but never become real.* They get shelved, or are never funded, or something else—a new shiny object—appears over the horizon.

Sometimes inaction is a matter of analysis paralysis or just plain old bureaucracy. To combat this at Amazon.com, Jeff Bezos has created a system of what he calls Type 1 and Type 2 decisions.

Type 1 decisions are irreversible turning points that top executives must be involved in.

Type 2 decisions, which are more common, are tactical calls that a business can reverse if it gets them wrong. These don't need the personal involvement of top leaders.

In 2015, Bezos wrote, "Type 2 decisions can and should be made quickly by high judgment individuals or small groups. As organizations get larger, there seems to be a tendency to use the heavy-weight Type 1 decision-making process on most decisions, including many Type 2 decisions." This leads to paralysis and inaction, as decisions are made too slowly.

3. Authenticity

Nobody likes a phony. Nobody! Unfortunately, many organizations have leaders who are disingenuous and, for the lack of a better term, "fake." Sadly, these leaders are under the impression that nobody notices, but people do. Their subordinates notice, their peers notice, and sometimes even sharp-eyed stock analysts notice.

Being slippery might be tempting as a short-term strategy to avoid confronting a problem. But in the long run, it's always a mistake. Sometimes even a legal one.

Life gets a lot better when you're willing to be self-aware, authentic, and honest. So many take a nosedive, and it's not surprising; in fact, it's often predictable. Your team is smart and they want to be respected. You want them to be authentic with you, and they expect that you will be authentic to them.

Adapt

H. G. Wells once wrote: "Adapt or perish, now as ever, is nature's inexorable imperative."

The renowned science fiction author may have been talking about the future of humanity, but winning leaders understand exactly what he was saying.

In business, change is the norm.

You either adapt or perish. Adaptation is the fueling force of evolutionary success. Moreover, adapting to massive deep changes in a time of disruptive innovation requires that organizations make adaptation a core competency.

You may assume that adaptation applies only to your product or service, but nothing is further from the truth. Adaptation can—and should—happen in every facet of your operations.

It can happen in:

- Human resources, as you keep pace with current trends that make your company the most desired place to work.
- Your supply chain, which is being revolutionized by the Internet of Things (IoT) and smart devices.
- Marketing, where the old strategies of interruption advertising are being supplanted by the power of social media and consumer connectivity.

- Manufacturing, where the emergence of 3-D printing and robotics threatens to disrupt the traditional assembly line.

Here are the three key attributes to being consistently adaptable and staying one step ahead.

1. Attention

It seems amazing, but many CEOs simply don't know how to pay attention.

Not necessarily to their paid consultants—that's their choice!—but to the marketplace and to their employees and stakeholders.

The winning leader is keenly aware of his or her environment, both internally and externally. You don't become aware and you don't learn by talking.

A good rule of thumb is that a leader should listen for three minutes for every one minute they spend talking.

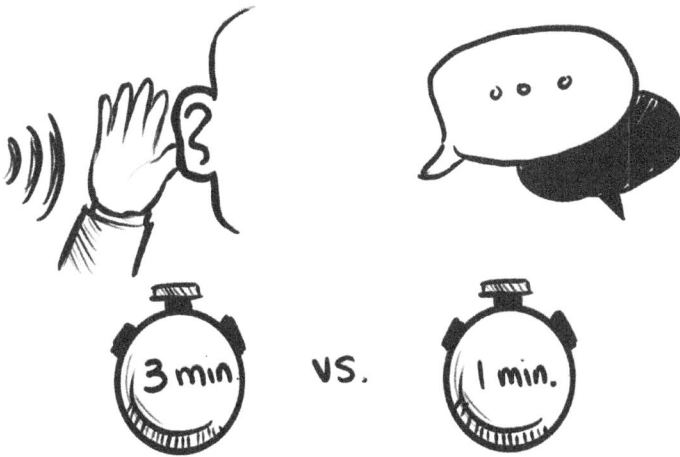

This view is shared by the cofounder of Google's successful career mentoring program. As Marguerite Ward wrote for CNBC, Jenny Blake, a career strategist who has helped more than a thousand Google employees climb the corporate ladder, advises that trying to solve a problem immediately usually does more harm than good.

"One of the biggest mistakes that I see managers making is immediately jumping in to give advice or trying to troubleshoot in the middle of a career conversation," Blake told CNBC, "rather than really asking open ended questions. Great leaders and managers make listening a priority. Not just any listening, but *active* listening."

2. Humility

Winning leaders have an unshakable set of core values: integrity, trust, honor, respect. These they never relinquish, and they take pride in them.

But—and not paradoxically—in their day-to-day activities they're not blinded by their own preconceptions or past successes. They're willing to learn from others and are open to new ideas.

As Jeanine Prime and Elizabeth Salib wrote in their *Harvard Business Review* article, "The Best Leaders Are Humble Leaders," a study by Catalyst revealed that humility is one of four key leadership factors necessary to create an environment where employees from different demographic backgrounds feel included. (And as we know, workplaces are increasingly comprised of a rich mix of demographics.) In a survey of more than 1,500 workers from Australia, China, Germany, India, Mexico, and the United States, researchers found that when employees observed selfless or altruistic behavior in their managers, they were more likely to report feeling included in their work teams. This was true for both women and men.

Altruistic or selfless behavior by leaders was characterized by:

• Acts of humility, such as learning from criticism and admitting mistakes.
• Empowering followers to learn and develop.
• Acts of courage, such as taking personal risks for the greater good.
• Holding employees responsible for results.

Among employees, the demonstration of humility by a superior serves to foster feelings of both individuality and engagement. Employees feel respected as individuals when they are recognized for the distinct talents and skills they bring to their teams. They feel engaged with the larger community when they share important commonalities with coworkers. Neither of these is possible when the leader is strutting through the halls like a peacock.

Based on their current research and an ongoing study of leadership development practices at Rockwell Automation, the authors noted that a "selfless leadership style" includes these action items:

Foster conversation, not debates. No one likes to be lectured, and no one wants to get into a pointless argument. As the leader, it's your job to calmly listen to all points of view, weigh the options, and clearly state your decision. Winning leaders are willing to set aside their own preconceived ideas when other better ideas are offered. Their egos aren't bruised!

Admit your mistakes and resolve to correct them. Employees know when you've made a mistake, and they know when you're trying to pass the buck. Do what you expect your employees to do: Fess up, take responsibility, say how you're going to fix the problem, and then do it.

Follow as much as you lead. By empowering others to lead, you'll not only facilitate the personal and professional development of your employees, but you'll model the act of considering different perspectives, a critical element to working effectively in diverse teams.

Embrace uncertainty. Since uncertainty and ambiguity are ubiquitous in today's business environment, then why fight them? Why not embrace them? If you humbly admit that you don't have all the answers, you make it possible for subordinates to step forward and offer solutions. By doing so, you also cultivate a sense of interdependence and reliance on each other as you collectively work through complex, open-ended problems.

An ancient Chinese proverb says, "The wise adapt themselves to circumstances, as water molds itself to the pitcher." So does the humble, winning leader!

3. Agility

It's not good when a leader acts like a deer caught in the headlights, frozen in place.

Chaotic changemakers are agile and move quickly to meet new challenges.

They put aside their immediate concerns to listen to those around them, and then they apply a set of specific skills and abilities to an externally perceived stimulus. They act upon that knowledge, attempting to help fulfill the needs of employees, superiors, and other stakeholders. Responsive leaders wield influence to solve problems for those around them, often before even being asked.

Above all, a chaotic changemaker strives to understand people and the operating context. They seek to understand what's really happening, as opposed to what they *want* to see. They recognize the constantly fluctuating nature of business and can quickly respond to new circumstances and challenges. They go one step further to anticipate challenges before they arise and quickly pivot to face them.

The opposite is the rigid leader, who when presented with an unfamiliar situation goes into defensive mode. Their brain automatically processes the problem as a threat, and their sympathetic nervous system shuts down, like a deer caught in the headlights.

Respect

Chaotic changemakers earn the respect of their employees and peers.

Notice the word "earn." It's not "deserve," or "are entitled to," or "should expect."

Any leader who struts around and expects to be respected just because of the nameplate on their door or the view from their corner office is making a big mistake.

Winning leaders earn respect every day and with every interaction they have with stakeholders. They never take it for granted.

Here are the three core areas in which winning leaders excel, earn the respect they deserve, and stay one step ahead.

1. Results

Chances are, you're demanding a lot from your team, and that's okay. But remember, your team is expecting *you* to deliver results. One of the best ways to drive leadership success is to focus on your own results as a leader so that you have the incredible benefits of sharing with your team that you expect nothing more from them than you expect from yourself.

Have you ever worked for a boss who demands a high workload but who shows up at 10 a.m. and leaves at 3 p.m.? (And don't say, "Oh, the boss is super-efficient." No. The boss is *lazy*.) We all have had such bosses, and we don't respect them. And chances are we won't work for them for long. As the leader, you need to "walk the walk" and set the example for your employees. When you know what's important to you, you're more likely to take action. Avoid distractions and busywork, and stick to tasks that keep you focused and help you move forward. Those around you will notice, and they'll make your work ethic part of their everyday behavior too.

Results—both that you achieve and you ask of others—must be clear, realistic, and measurable. Don't play games with your employees by being vague or by moving the goalposts. Setting SMART (specific, measurable, achievable, relevant, and timely) goals can help you evaluate the goals you wish to set. Put a system in place to help you measure the goals of both yourself and your employees, and keep the organization on track.

2. Responsiveness

Have you ever worked with someone who, in response to your question, often said, "I'll get back to you on that"—and they never did? I'll bet you felt not only annoyed because you didn't get an answer, but also personally disrespected—as if you were insignificant and your work didn't matter.

One particularly insidious nonresponsive archetype is called *the bunker*. The bunker believes that he or she can hide out, and because they're not interacting, nothing bad will happen. Of course they're wrong. Prompt and clear responses to input from subordinates or peers are vital to building a culture of respect. We owe accurate and timely and complete responses to our team, and winning leaders make it part of their leadership belief system and behaviors.

3. Reality

Chaotic changemakers have their feet planted squarely on the ground of *reality*. These lucid leaders understand their business, the markets, and competition, all in the timeline that it takes to accomplish specific tasks and duties. They have invested in understanding their business, teams, and industrial ecosystems so well that they always know exactly what's possible and what's not. This is not a small matter. Team members who are expected to do more than they can really do or to achieve something that is simply not real leave organizations quickly.

Chaotic changemakers know the difference between what's possible and what's impossible.

It's unrealistic to expect a complex piece of software to be fully functional in its first iteration.

It's unrealistic to think that customers will be overjoyed when you roll out a mediocre product.

It's unrealistic to think your people can make miracles happen without adequate resources and backing from the corner office.

To be realistic, you first need to be lucid, which means being self-aware as well as cognizant of the external environment. You need to encourage your team to be lucid, and you need to engage their insights and be deeply connected to what human beings are capable of producing.

Take Action!

✗ **Changemakers need to have both depth and breadth.** They need the Value Leadership Model. The Value Leadership Model behaviors are Inspire, Connect, Adapt, and Respect.

   Each behavior has three attributes. All are equally important:

   Inspire = Vision, Value, Attainability.

   Connect = Access, Action, Authenticity.

   Adapt = Attention, Humility, Agility.

   Respect = Results, Responsiveness, Reality.

✗ **Create a powerful inspirational beacon that guides team members and draws them closer.** Is your organizational vision statement both inspirational and aspirational, and does it create a mental image of the future state that the organization wishes to achieve? The leader at the top must understand, endorse, and promulgate the organizational vision, and ensure that it adapts to rapidly changing conditions, both internal and external.

✗ **Proactively reach out to engage team members in a way that is relevant to their individual archetypes, goals, and responsibilities.** With leadership comes responsibility! Winning leaders earn the respect of their employees and peers every day and with every interaction.

   We've always known that leadership required clarity and everything that's mentioned in this chapter. We may just want to put one sentence in the talks about in a time of chaotic change. We need to get 100 percent stakeholder adoption to the mission in order to compete in the marketplace of chaotic change.

CHAPTER 3

Directors and Investors

In business we talk about "managing down," which means leading your subordinates at whatever level they are and you are. If you're a shift manager, you lead your immediate reports, which could be as few as a handful of people. If you're the CEO, you lead everyone who's lower on the organizational chart than you, which could include thousands of people.

We also talk about "managing up," which means taking control of your relationship with your superiors. Generally, it means giving your boss exactly what he or she expects. You need to understand what makes your boss tick (and what ticks them off) and know how to anticipate their needs. Sometimes friction is inevitable, but knowing the right way to bring a problem to your boss can help you navigate sticky situations. There will be times when you disagree with them, and that's fine, as long as you're careful to disagree in a respectful, productive way.

It means being the most effective subordinate you can be, creating value for your boss and your company.

As an executive leader of your organization, you may have multiple bosses—the members of your board of directors and your investors or owners. You might say that in dealing with these bosses, you need to be the "changemaker subordinate." You need to manage up by doing your best to create a culture of chaotic change for everyone above you.

Your job may depend on your success at doing this.

The Corporate Pressure Cooker

While every relationship in an organization is a two-way street, it's exceptionally so when it's between the CEO and the investors and board members. The stakes can be very high, and both sides need to see and accept a common set of validated facts.

Given the enormous pressure that can be put on the CEO and top executives to produce a steady flow of good news, this is often difficult. In recent decades, the compensation and job evaluation of a CEO have become increasingly tied to the stock price. When the stock price goes up, the CEO is a hero. When the stock price slumps, regardless of the reasons, the CEO is a loser. Job performance is also tied to quarterly earnings reports. Investors of the non–Warren Buffett variety are increasingly impatient and view a quarterly loss, for whatever reason, as a failure.

This book could be filled with lurid stories of scandals in which CEOs and top executives have lied to investors and board members about their company's profitability. But if you Google the worst corporate financial scandal and the worst example of the failure of a board to provide lucid leadership oversight of an executive team, one name—which has become the stuff of legend—always appears.

Enron.

Like Benedict Arnold or Pearl Harbor, the name has become a part of popular lore. People may not know the details, but they know that something *bad* happened.

A Houston-based global energy company that grew at a phenomenal rate between 1996 and 2000, Enron reported an increase in sales from $13.3 billion to $100.8 billion. In August 2000, Enron shares reached their all-time high of $90 dollars per share. Investors were delighted.

The company appeared to be brilliantly managed—but appearances were deceiving.

On December 2, 2001, Enron declared bankruptcy. The next day the company laid off 4,000 workers. (An employee later revealed that Enron had paid $55 million in retention bonuses to top managers and executives just before going belly-up.)

What happened?

In an ethical company, the financial results reported by managers to investors are verified by the company's independent auditor. The auditor's work is overseen by the company's board of directors, which should be made up of objective, lucid people who have the best interests of investors at heart. This three-pronged system of management, auditors, and directors is designed to assure investors, who are putting their money at risk when they buy stock, that they are getting an accurate picture of the company's performance.

The chief architects of the scandal were the top executives: Kenneth Lay, Enron's CEO and chair; Jeffrey Skilling, head of Enron Finance Corp.; and Andrew Fastow, chief financial officer (CFO). When the company started to pile up debt and toxic assets, rather than fix the problem, Fastow orchestrated a scheme to use off-balance-sheet special purpose vehicles (SPVs), also known as special purposes entities (SPEs), to hide the growing cancer.

You may ask, isn't the outside auditor supposed to inspect the books and reject unethical practices?

Yes, it is. Enron's auditor was the accounting giant Arthur Andersen, and specifically its Houston office. During 2000, Arthur Andersen earned $25 million in audit fees and $27 million in consulting fees from Enron. The auditor's shoddy reports and lack of expertise in reviewing Enron's derivatives, revenue statements, special entities, and other accounting practices were the result of the firm's priority being fee collection rather than professional rigor.

How about the Enron board of directors? Why didn't they see what was going on and step in to curb the madness?

On paper, Enron had a respected board of directors comprised predominantly of outsiders with significant ownership stakes and a capable audit committee. Enron directors were highly paid for their services and financially tied to the company. In 2001, the average Enron director was paid nearly $400,000 in cash and stock, using the value of Enron stock on the date of its annual meeting.

Four hundred thousand dollars is the same salary paid to the president of the United States. And unlike that of an Enron director, the presidency is a full-time job.

The board members were easily manipulated by Fastow.

In the end, Enron investors lost $74 billion in the four years leading up to its bankruptcy, and its employees lost billions in pension benefits. Lay, Skilling, Fastow, and others were sentenced to prison.

While Enron was perhaps the most spectacular case of corporate fraud, the reality is that executive leaders in companies large and small face enormous pressure to perform. (In all fairness, sometimes the CEO is the primary instigator, as we saw in the case of Elizabeth Holmes, chairwoman and CEO of Theranos, who was convicted in 2022 of four counts of fraud that carried a maximum sentence of 20 years behind bars.)

As the chaotic changemaker, you need to be especially on guard against the pressure that may be exerted on you by the board and investors to deliver the news they want to hear rather than the news that's real. You also need to be careful that you—either deliberately or without thinking—do not put pressure on your CFO to do the same.

You might say, "I'm a lucid leader. I play it straight and I don't like fabricated numbers. But certain board members are hardliners. They don't care. They want what they want, and if they're not happy, they may fire me."

Yes, it's true that some boards and investors are single-minded and don't understand the realities of our hyper-competitive world. But whose fault is that? If you're a founder of your company, you have some control over whom you accept as an investor or board member. You can be lucid and you can demand they be the same. In the first years of Amazon, Jeff Bezos had a handful of investors, including his parents; in all, about 20 investors ponied up checks of about $50,000 apiece, for a total seed stage investment of around $1 million. In exchange, those combined investors got about 20 percent of the company. In June 1996, Amazon raised an $8 million Series A from Kleiner Perkins, its only VC investment before going public in May 1997. The IPO raised $54 million, giving the company a market value of $438 million.

The company grew incredibly quickly, and its investors didn't mind that it didn't report a profit until the fourth quarter of 2001. Bezos, his investors, and his board were all equally lucid about the company's growth strategy.

If you're brought into an existing company from the outside, and particularly a company that has a powerful and well entrenched board, then you need to be super-lucid about what you're getting into. Now *you're* the candidate being interviewed for the job, and it's incredibly tempting to tell the board members what you think they want to hear.

Two-Way Lucidity

Unless you own and control your company 100 percent (which would make you a super-sole proprietor), while you may be the chief executive, you're going to be reporting to your board and investors or owners. You can be as lucid as you want, but if the lucidity isn't shared, or if your counterparts don't see things the same way you do, then there are going to be crossed signals and conflict. When that happens and it's serious, it's the CEO who gets shown the door.

But the ball is always in your court. You control what you tell your board about the performance of the company.

You may be tempted to tell them only good news. You may think, "Why not make them happy? If there's bad news, we can fix it before the next board meeting." This is extremely dangerous, because you're being dishonest and because the problem you're hiding may only get worse.

You may be tempted to present a litany of complaints and doom-and-gloom, thinking that every little victory will then be celebrated even more. This is equally dangerous because the board may conclude that you're incompetent.

Two-way lucidity—the exchange of truth—builds trust and strengthens your relationship to your board. You need them to believe you, and you also need them to pitch in and help when necessary.

Here's how to ensure your relationship with your board is based on reality and mutual respect.

1. Remember the Lucid Leader Manifesto:

The Lucid Leader:

- Is honest about themselves.
- Sees the world as it is, without preconceptions.
- Pursues the mission of the organization.
- Strives to make a positive impact on the world.

Every word you say to your board needs to be in alignment with these principles. Be honest, don't put a "spin" on bad news, and treat your board members like they're intelligent, caring people. If you have one board member who's a "problem child," just bite your tongue and say as little as possible. Remember: If a board member seems obnoxious to you, it's guaranteed he or she is obnoxious to the other board members as well.

2. **Tell them the most important news.** Board meetings can easily become bogged down in minutiae. The board chair needs to control this tendency, but sometimes they don't. Individual investors and board members probably don't know any more about your company and your market than you. As the CEO, it's your job to make sure your board promptly and succinctly knows about important items. If a website defect harms your relationship with customers, or a competitor makes an out-of-nowhere acquisition offer, tell your board chair ASAP, not next month. It doesn't build trust when they find out from elsewhere.

3. **Tell them your biggest challenges.** Board members do not need to know about the everyday problems that you get paid to fix. They need to know about serious issues that their expertise might be able to solve. Ask them for advice and come up with an action plan together of how you will try to solve it. Outside of regular meetings, figure out what you want advice on, then seek out the board member who is most qualified to offer advice on that topic.

4. **Ask them to plan ahead.** One of the most important functions of a board is long-range planning. Share with them your vision and ask them for theirs. Be sure to iron out any differences while you're still in the talking phase. Aside from policing and oversight functions, the primary place directors can add value is in offering a different perspective on the competitive environment and the changes in that environment.

5. **Ensure you're on the same page.** Start with a conversation about shared values between yourself and your board members that includes such expectations as open-mindedness, accountability, financial stewardship, transparency, and confidentiality.

6. **Don't be the board chair.** It's become common in the United States for the CEO to also serve as the board chair. This is a bad idea. The same person can't do both jobs. Because an important part of the board's job is to evaluate the performance of the

CEO from the point of view of the shareholders, it's a conflict of interest and the immediate result will be a loss of lucidity.

How to Choose Your Investors

Choose your investors? To many entrepreneurs, this may seem like a crazy idea! They would think, "Hey, if some Silicon Valley venture capitalist wanted to drop a couple of million dollars into my company, I'd be a fool to say 'no'!"

Not so fast!

The lucid leader knows that the type and size of investments, and the conditions under which they are offered, can either save or sink a company.

In a perfect world, every business could self-finance, and the owners wouldn't need to worry about repaying a loan or dealing with equity investors. But in the real world, businesses often need cash, usually to expand, continue research, or pay for a big purchase order that will produce revenue down the road.

Three types of funding are possible.

1. **Debt.** This means you take out a cash loan from a bank or private lender. You promise to repay the loan over a certain period of time. The lender is not entitled to any ownership percentage—they take their profit from the interest they charge or from a royalty on each item sold. If you don't repay the loan, they can sue you or foreclose on your company.

   As long as you make your payments on time, the financier has no control over your business. The interest you pay is tax-deductible. Once you've repaid the loan, your relationship with the lender ends.

   It's easy to forecast debt expenses because loan payments do not fluctuate over time. From a perspective of pure risk, lower ratios (0.4 or lower) of income to debt are considered better.

Since the interest on a debt must be paid regardless of business profitability, if cash flow dries up, even temporarily, too much debt may compromise the business. Companies unable to service their own debt may be forced to sell off assets or declare bankruptcy.

2. **Equity.** Here, the investor provides funds in exchange for partial ownership of the company. They get paid back either through profit distributions or when they sell their shares.

   The only way to remove investors is to buy them out, but that's likely to be more expensive than the money they originally gave you.

   The more cash the investor provides, the more equity they want. If the investor ends up owning 51 percent of the company, you are now effectively their employee.

3. **Crowdfunding.** For smaller ventures, crowdfunding is a quasi-charitable method with a multitude of variations. In 1885, the Statue of Liberty was completed thanks to a crowdfunding campaign spearheaded by publisher Joseph Pulitzer. While many of today's crowdfunding campaigns are for projects related to blockchain technology, one of the biggest crowdfunding campaigns is the U-Haul Investors Club, in which participants buy "U-Notes" to invest in U-Haul assets such as trucks and trailers. As of August 2022, the amount invested was over $194 million. AMERCO, the parent company, makes it clear that when you purchase a U-Note, you are lending money to AMERCO, which repays that loan by making scheduled repayments to you through the U-Haul Investors Club.

For the lucid leader, when choosing the type of funding you need, there are two cardinal rules: Never give up majority stake in your company, and never take on debt that you're not certain you can repay.

Lucid Business Reports

Just as the lucid leader must *perceive* the world with objectivity, so must he or she *communicate* with transparency in all business matters. This includes all communications with investors, regulators, and the general public. By being transparent, the leader will build and reinforce his or her reputation for honesty.

Quarterly earnings reports are a common area where deceptive leaders seek to bury bad news while finding ready audiences for half-truths. One trick that unethical leaders frequently use is to release negative information after the close of the market on a Friday afternoon, especially when heading into a holiday weekend. Or a company might announce their lackluster earnings after hours when there's typically a lower level of investor attention, or on a day when hundreds of other companies are reporting and the analysts are distracted.

The chaotic changemaker knows that for investors, *less information* means *less certainty* and *higher risk*. When financial statements are not transparent, investors can't accurately determine a company's real fundamentals and level of risk. Complex, vaguely worded reports may also obscure the company's debt level, which is what Enron did. If a company hides its debt, investors can't estimate their exposure to bankruptcy risk.

A company's prospects for growth are related to how it invests. It's difficult to evaluate a company's investment performance if its investments are tucked away in holding companies and hidden from view.

Take Action!

✗ As the CEO, you're not at the very top of the organizational pyramid. You're occupying a spot on the circle of accountability. Your skill as a chaotic changemaker must impact your investors and board members as well as your subordinates.

✗ Maintain close communication with the board chair and other board members who are active. Encourage them to be lucid as well, and try to tamp down unrealistic or "pie in the sky" notions.

✗ Focus the board on long-range planning, not short-term micromanaging.

✗ Lucidity is a two-way street. See the world clearly, and let the world see *you* clearly.

CHAPTER 4

Make Innovation REAL

To sustain healthy growth, innovation—which we define as *the creation of new value that serves your organization's mission and customer*—must be woven into the fabric of the organization and sustained over time. In a time of chaotic change, your organization needs to innovate consistently and methodically. There are many ways to do it, and you need to find the methods that work best for your company.

Luckily, human beings are born innovators—so you're already halfway there. And you probably have pockets of innovation happening right now in your company. That's good!

But you want consistency and quality. As a leader, it's your job to set up a *system* to take your employees' sparks and collect them, evaluate them, and treat them as potential assets. To exploit their profit potential, selected innovations need to be transformed from *theory* to *reality*.

This transformation will not happen on its own or by accident. Profiting from innovation takes a consistent and structured effort.

Unfortunately, too many sparks of innovation are ignored, overlooked, or allowed to get cold and die. Sometimes they seem attractive when first proposed, but there's no will or process to take the first small spark and turn it into real energy.

Success in our time of chaotic change begins with leadership. Without effective leadership, no organization can hope to survive, much less thrive. Here are the key concepts and lessons that will help you strengthen your own leadership capabilities and instill the power of leadership in every corner of your organization.

The Make Innovation REAL system is very simple.

REAL = Review + Encourage + Act + Lead

Let's examine it in detail.

Review

Leaders who become enamored with the idea of innovation often plunge ahead pell-mell without ensuring that their organization is ready to embrace a culture of sustained innovation. If they do this, they run the risk that the spark will fall on damp wood and fail to ignite. This is not the outcome you want.

Without a robust innovation operating system, things like hosting quarterly hackathons, buying Ping-Pong tables, making an unused room into an "innovation lab," installing whiteboards, and making jeans acceptable work attire are nothing more than quick-fix solutions. Leaders sometimes see these measures and become convinced they're ready for innovation when, in fact, the critical ingredients for sustained, profitable innovation are missing.

The Two Critical Questions

The first question to ask is, "As the leader, am I personally ready to embrace new ideas that can be shown to have merit and are worth investing in?"

The answer needs to be, "Yes, I'm ready."

The second question is, "Are our employees in a psychological place where they will trust leadership to treat their new ideas with respect? Will they respond to our call for innovation and new ideas, or think it's a ploy?"

Again, the answer needs to be, "Yes, they're ready."

Fractional, piecemeal innovation initiatives aren't enough. To fully leverage the benefits of chaotic change, you need to review the current state of the business before designing an innovation operating system touching every corner of the organization and providing a step-by-step template aimed at building market-grabbing power. This should be done with input and ownership from all levels and across all business units.

Your *innovation mandate*—as we call it—needs a clear direction. Set goals, both large and small, long term and short term. For any type of innovation—from saving time in a process to improving a product—identify what you want to achieve, how long you have to get it accomplished, and what constitutes success.

Encourage

After an organization builds out a credible innovation strategy, leaders need to approve, endorse, and fund the systems and tools necessary for successful execution. In other words, the formation and acceptance of a theory must be followed by clear directives supported by leadership.

Leaders are the key players in fostering a culture of innovation, including modeling behaviors to ensure that the walk matches the talk. Because failure is a built-in and necessary feature of innovation, this can often mean showing support for untested, new, or disruptive ideas. It's imperative that leaders consistently communicate the vision of innovation so that no one misses the message.

An effective innovation mandate comprises a range of activities and systems that encourage your team members to participate in the game of innovation. (And yes, it is like a game, with really good prizes for the winners!) These processes are designed to increase the volume of ideas that go into the *innovation pipeline*, which is your system for identifying and developing new ideas. Remember, innovation is a high-volume, low-yield proposition. Put more simply, *failure is a fundamental part of innovation*. For every ten ideas your people propose, perhaps one will survive and show itself to be profitable. That's to be expected!

The innovation pipeline is no different from the standard project development process that you already know. In any project development cycle, the steps are:

1. Brainstorming
2. Collaborating
3. Planning
4. Implementation
5. Evaluation
6. Completion (unless it's ongoing)

During the brainstorming phase, a premium is placed on *ideas*. Good ideas, bad ideas, crazy ideas—as you know, every project development expert on earth will tell you that during this important first phase, every participant must have permission to offer any far-fetched notion. When everyone's had a chance to contribute, the inventory of ideas is discussed and the ones that are promising are retained while the impractical ones are discarded, with no judgment passed on the people who offered them.

A culture of innovation is no different. In the earliest phase, *quantity equals quality*. The more new ideas, the better!

Act

Following reviewing and encouraging, leaders and employees must *act* upon their stated innovation strategy. The goal is to identify, evaluate, nurture, and underwrite new ideas from all sources. Keeping the innovation pipeline full needs to be a priority. If the company is big enough, this may require an innovation director or department to manage and track the innovations that will come streaming in.

A well-designed *employee suggestion program*, supported by organizational commitment, leadership clarity, and ongoing communication, can positively impact your employee motivation and enthusiasm, your innovation pipeline, and ultimately your bottom line. You can also schedule departmental brainstorming sessions or solicit ideas during a few minutes of your weekly staff meeting. Some companies set aside one day a month for a lunch meeting at which every employee is asked to submit at least one idea.

Innovation hackathons are an effective way of targeting specific enterprise goals. Their use will significantly improve your return on strategies and stated enterprise priorities. The key is to innovate with your internal stakeholders, as they will provide you with amazing insights that can yield significant enterprise benefit. Remember to create a culture that demonstrates a safe place to innovate by acknowledging the potential of failure and letting stakeholders know that there are no repercussions for trying new ideas.

If your people need a structure that will help them get started, digital innovation platforms such as Qmarkets, which enable leaders to tap into the collective intelligence of employees, partners, and customers to find the best ideas and make the right decisions, can help engage stakeholders to participate in well-defined innovation challenges. Be sure to prepare by conducting a comprehensive innovation gap analysis, and you'll need a high-level innovation mission prior to identifying any technology tools. Innovation is a people-powered process, and while it can be optimized through a range of technology tools, you should use them with great care and thoughtfulness.

Respond Decisively

Having assembled an inventory of new ideas—or even better, having a steady stream—the next action step is to evaluate each one and then take one of these decisive actions:

1. Accept it.
2. Reject it.
3. Send it back for more study, which may include funding it.

Remember, to the employee who submits an idea to "the bosses," nothing—I repeat, *nothing*—is more disheartening than receiving no reply. Every idea should be acknowledged within twenty-four hours, even to just say, "Thank you, we appreciate your idea." If possible, the employee should be notified as to the dispensation of their idea—yes, no, or further study.

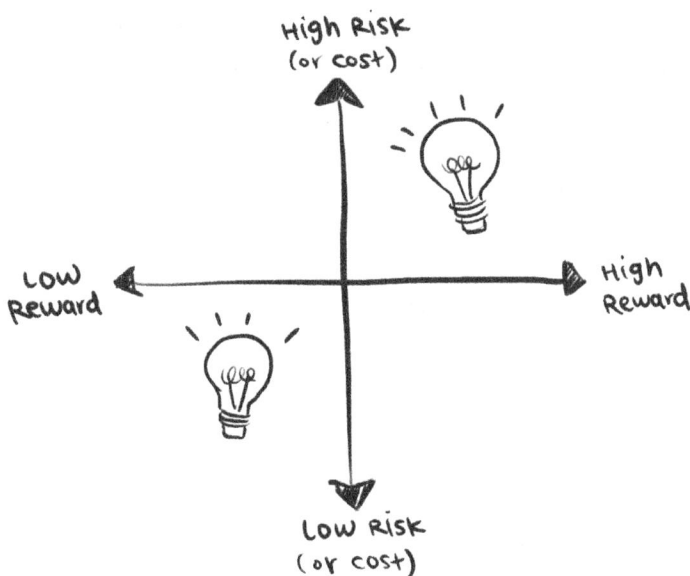

Success in our time of chaotic change begins with leadership. Without effective leadership, no organization can hope to survive, much less thrive.

Here are the key concepts and lessons that will help you strengthen your own leadership capabilities and instill the power of leadership in every corner of your organization.

In your inventory of new ideas, you'll quickly see that they tend to fall along a spectrum of risk (or cost) and reward. If you created a coordinate grid, you'd plot the vertical *x*-axis with low risk (or cost) to high risk (or cost), and the horizontal *y*-axis with low reward to high reward. At the lower left corner you'd find the ideas that were low risk and low reward. In the upper right corner would be the ideas that are both high risk and high reward.

Risk and Reward Matrix

In this example, Idea 1 is low risk/cost and low reward. It may not look like much, but if this idea were one of a constant stream of low-risk, low-reward ideas in a program of continuous incremental innovation (*kaizen*), then it would be valuable. Remember, as Katsuaki Watanabe of Toyota said, if you keep innovating like that for seventy years, you'll have a revolution!

Idea 2 is low risk/cost and high reward. What are you waiting for? Grab it and go!

Ideas 3 and 4 are higher risk/cost with low expected reward. You may want to pass on them.

Idea 5 is high risk/cost and high reward. These are the breathtaking innovative gambles that make headlines, like a movie studio wagering hundreds of millions of dollars on a summertime superhero blockbuster. If you gamble big, then you win—or lose—big.

On our risk/reward graph, it would seem obvious that Idea 2, which was both low risk and high reward, would be highly desirable! But amazingly, even ideas that have landed in this sweet spot have been rejected by leaders who didn't recognize or value innovation.

There are so many juicy examples . . . here are just a few:

- "Who the hell wants to hear actors talk?" *Harry Warner, a founder of Warner Brothers movie studio, said this in 1927, when movies were silent.*
- "There is no reason anyone would want a computer in their home." *Ken Olsen, president, chairman, and founder of Digital Equipment Corp., said this in 1977.*
- "So we went to Atari and said, 'Hey, we've got this amazing thing, even built with some of your parts, and what do you think about funding us? Or we'll give it to you. We just want to do it. Pay our salary, we'll come work for you.' And they said, 'No.' So then we went to Hewlett-Packard, and they said, 'Hey we don't need you. You haven't got through college yet." *Apple Computer Inc. founder Steve Jobs shared this about his early attempts to interest big tech companies in his personal computer.*
- "On June 26, 2008, our friend Michael Seibel introduced us to seven prominent investors in Silicon Valley. We were attempting to raise $150,000 at a $1.5M valuation. That means for $150,000 you could have bought 10 percent of Airbnb. Below you will see five rejections. The other two did not reply." *Brian Chesky, cofounder and CEO of Airbnb, wrote this in a blog post.*

Yes, even smart people are fallible. As a leader, you need to be open to new ideas, know how to evaluate the risk and reward of an idea, and take action when necessary.

Remember, you can't focus exclusively on risk! Over the years, consultants and leaders have created many types of processes for innovation evaluation and management. Most are incredibly complicated and definitely risk centered. In fact, the overwhelming majority of organizations use systems that are virtually exclusively centered around *risk management.* This approach is like playing the game not to lose rather than playing the game to win.

The innovations that come out of risk-centric evaluation and management processes are rarely disruptive or breakthrough innovations that have big opportunities. These types of processes typically incubate incremental improvements. The best way to look at the innovation mandate is that it's a stock portfolio comprised of a range of high-risk, high-reward innovations and lower-risk, lower-reward innovations. But you know the rule that risk and reward increase together, so if you want the big hit, you're likely going to take a big risk. Afraid of taking a big risk likely means you will see only incremental innovations.

Lead

The last letter in Make Innovation REAL is "L," as in *lead*.

According to the Center for Creative Leadership, "Studies have shown that 20 to 67 percent of the variance on measures of the climate for creativity in organizations is directly attributable to *leadership behavior*. What this means is that leaders must act in ways that promote and support organizational innovation." Day in and day out, innovation requires sustained and powerful leadership. The best innovation strategy in the world will fail without committed leadership. Innovation requires savvy leaders who possess a core competency around innovation and a commitment to make innovation part of the enterprise of DNA and reap the rewards, day after day, quarter after quarter, year after year.

The Center for Creative Leadership describes the three tasks of leadership as setting direction, creating alignment, and building commitment. When these core tasks are centered around innovation, organizations become more innovative and more productive.

Your innovation mandate needs to be capable of transforming ideas into reality. Your new process improvement, technology, marketing innovation, or other enterprise innovation needs to be plugged into your organization's product lineup or everyday operations so its worth can be proven or disproven.

This requires the endorsement and support of leaders, because they are the ones who typically control the allocation of resources in the form of both money and time.

Innovation—whether it's a new invention, new product, or new business process—represents *change*. A change from the way things are being done now. If you know anything about people, which I'm sure you do, then you know that most people in their day-to-day work don't seek out changes in their routine.

Innovation Champion

Whether it's for a particular project or to manage a culture of innovation, you may need an *innovation champion*. This is someone who, when necessary, can get the endorsement of leaders when it's time to put an idea into practice.

An innovation champion is passionate about making innovation thrive within their organization. Champions may not necessarily be the top idea generators or creative geniuses; rather, they are the inspirers, facilitators, and connectors. A critical player in the success of the deployment, the innovation champion acts well beyond the initial market launch. They complement the role of the project development leader. During successive launches, they progressively develop a keen understanding of the benefits valued by customers and employees alike, guiding those who adopt the innovation by providing them with logistical, technical, and economic support.

Take Action!

✗ **Know your company.** Review your capacity for innovation versus where you should be. Get input from a wide range of stakeholders. Be honest: Are you sitting on your hands because your market seems stable and your customers happy? You should not feel so comfortable! Industry disruption is happening more quickly and more deeply than ever, and if you fail to create a culture of innovation, you'll find your market position eroding.

✗ **Design your own REAL innovation mandate.** While this book provides guidelines and key insights, no two companies are alike. An innovation mandate that works for one company isn't going to work for another—or for your company. You need to identify and nurture your organization's innovation strengths to match your mission and financial goals.

✗ **Launch your minimally viable organizational transformation to become an innovation leader.** Don't wait until it's perfect! Remember, innovation is a process that will become an integral part of your company, not a short-term project. Get it up and running. Evaluate it and make adjustments. Get feedback from stakeholders. If an idea looks promising, fund it and demand results. Spread the news when you have a success. Don't worry about crazy ideas—let them go and focus on the good ones. Make your employees a vital part of the innovation pipeline. Don't stop!

CHAPTER 5

The Rise of
Artificial Intelligence

This chapter is about the limit, or horizon, of human knowledge, and how it's been exploded by the rise of artificial intelligence.

The term "knowledge horizon" refers to the scope of knowledge attainable by leaders in the execution of their daily responsibilities. That scope is expanding at an increasingly faster rate, just as we now know the universe itself is expanding at a faster rate (which is baffling to astrophysicists because it's violating the laws of gravity—but that's a subject for a different book).

If you think about it, for thousands of years, a leader had three ways of knowing what was going on in their organization, such as an empire, an army battalion, a religious group, or a feudal estate. These three ways were very simple.

1. Direct personal observation of an event. This was uncommon, because no leader can be everywhere at one time. The leader's range of knowledge was literally the visible horizon.
2. A report of an event delivered orally by a trusted person. This was most common—every leader had a network of subordinates

and spies. This capability extended the leader's range of knowledge over the visible horizon.

3. A written communication, such as a letter. This was fairly common, but letters took time to physically deliver. In theory, with written letters, the range of knowledge could extend around the world.

Leaping ahead to the late twentieth century, business leaders had the telephone (with "long distance" calls abroad) and the newfangled fax machine, which could transmit documents over physical phone lines.

With accelerating technology, the twenty-first century brought instant digital communications, both voice and print, followed by video services such as Skype, which was quickly supplanted by Zoom. (Speaking of chaotic change, the dizzying rise and fall of Skype is a good case study!)

Machine learning allowed leaders to take and quickly process large quantities of *known data* find *future* patterns and trends. For example, DNA analysis requires vast amounts of storage space and processing speed, both of which emerged in the first decade of the twenty-first century, making sequencing fast and affordable.

These advances allowed leaders to become more lucid in regard to their ability to get information about events and developments occurring at a distance, as well as complex events happening anywhere.

Artificial Intelligence

By the third decade of the century, AI had emerged as a potent tool to expand the knowledge horizon of leaders. It also quickly became something fearsome, with more than one prognosticator calling AI an "existential threat to human existence." To be fair, we've seen this response before: for example, in the mid-nineteenth century, as railroad trains began to reach the fantastical speeds of 30 miles an hour and more, many learned people asserted that such high rates would lead to mental distress and violent internal injuries.

In general, AI systems use *machine learning* to recognize patterns in data in much the same way as the human brain learns. If you show a child 1,000 pictures of varieties of apples, the child will learn to recognize the characteristics of an apple. Then if you show a child a picture of a lemon, despite some general commonalities with an apple, the child will be able to say, "That's not an apple. It's something different."

In this case, the pictures of the apples are what AI scientists call *labeled training data*.

By ingesting large amounts of labeled training data and analyzing the data for correlations and patterns, the AI-capable computer can make predictions about future states. A chatbot that is fed examples of text chats and rules of grammar can learn to produce lifelike audio exchanges with people, and algorithms such as Chat GPT can produce original written text.

Without a doubt, AI is a major driver of chaotic change. You might even say it's one definition of chaotic change, and it's having a profound impact. A research study by Infosys that polled over 1,000 global C-level executives at large organizations across seven markets revealed that AI is becoming a core aspect of business strategy. Broad adoption of AI is impacting every aspect of the way leaders do their jobs, including the way they drive innovation and compete, inspire teams, recruit and train, and apply AI and human power together to achieve their vision for the company. Forty-five percent of IT decision makers report improved process performance from AI, and 40 percent report productivity gains due to IT time spent on higher-value innovative work.

A primary factor driving ROI from AI is the presence of a clearly defined strategy. Eighty percent of respondents who said they've seen at least some measurable benefits from AI also noted their organization had a defined strategy for deploying AI.

Chaotic changemakers evolve quickly in a disruptive and dynamic environment. And to bridge the gap between old and new business and work realities, they champion change throughout the organization. In the Infosys poll, 76 percent of all respondents were either confident or extremely confident that the senior leaders of their organization understand and promote the positive aspects of AI.

AI is on track to destroy many established businesses and even industry sectors. The upside is that it will create completely new businesses and industry sectors that will provide the human connection between the machine and the human. It has been said that AI itself will not replace a business, but it will replace a business that *doesn't use AI*. Like the internet, AI is a double-edge sword that is both creative and destructive.

A Computer on the Board of Directors

In 2014, a Hong Kong venture capital fund announced it had appointed a computer to its board, with voting power.

Strictly speaking, Deep Knowledge Ventures (DKV), a firm that focuses on age-related disease drugs and regenerative medicine projects, claimed to have appointed an algorithm called VITAL, which could make investment recommendations about life sciences firms by analyzing large amounts of data.

"[The goal] is actually to draw attention developing it as an independent decision maker," DKV's Charles Groome told *Business Insider*.

Dmitry Kaminskiy, managing partner of DKV, said that the fund would have gone under without VITAL because it would have invested in "overhyped projects." VITAL, which stands for Validating Investment

Tool for Advancing Life Sciences, helped the board to make more logical decisions, he said.

Speaking of "overhyped," when the announcement received widespread press coverage, industry professionals responded with a healthy dose of skepticism. Michael Osborne, an associate professor in machine learning at the University of Oxford, said, "Essentially, all I think they're doing is using the predictions made by this algorithm as kind of a starting point for discussion on the board, which I think is a totally reasonable thing to do, but I think it's a bit of a gimmick to call that an actual board member."

Another small detail is that according to Hong Kong law, board members must be actual human beings.

But the point was made: AI is increasingly being used to expand the knowledge horizon of lucid leaders, giving them insights that are deep and can be produced quickly. More effectively than the human mind? Legendary investor Warren Buffett might have a thing or two to say about that. But not every human being is Warren Buffett.

This leads to an interesting point. Humans have very complex skill sets. In terms of evolution, some of those skills have been a part of our operating systems for millions of years, such as facial and voice recognition, navigating through space, catching a ball, judging people's motivations, setting future goals, and all the skills that we use for attention, visualization, perception, movement, socializing. These are "easy" skills that we can do without conscious thought.

In addition, we possess skills that we've developed in more recent times: mathematics, engineering, scientific reasoning. These require conscious thought. They do not come easily to us—they are the "hard" skills.

As it turns out, AI algorithms are good at these modern "hard" skills, and not so good at the ancient "easy" skills.

This has led to what is called Moravec's paradox. Articulated by robotics and computer scientists Hans Moravec, Rodney Brooks, Marvin Minsky, and others in the 1980s, it's the observation that, contrary to traditional assumptions, reasoning requires very little computational resources, but sensorimotor and perception skills require enormous amounts. As Moravec wrote in 1988, "It is comparatively easy to make computers exhibit adult level performance on intelligence tests or playing checkers, and difficult or impossible to give them the skills of a one-year-old when it comes to perception and mobility."

In the development of AI, computer scientists were initially successful at writing programs that used logic, played complex games including checkers and chess, and solved algebra and geometry problems. These tasks are difficult for most people and are considered a sign of intelligence. Many researchers assumed that having made progress solving the problems that were "hard" for humans, the "easy" problems of vision and common-sense reasoning would soon follow. They were mistaken. Such problems are extremely difficult to solve, but we humans, with millions of years of practice, make them appear easy.

In business, we must make do with the skills and capabilities we possess while seeking help with the tasks that are more difficult—and for many chaotic changemakers, an AI algorithm can provide that extra competitive edge.

Worker Productivity Tracking

Chaotic change is a double-edged sword. New technology brings new benefits and new challenges. It can make life easier and more productive while raising ethical and even legal questions.

Worker productivity has always been subject to measurement. Since the earliest days of the Industrial Revolution, assembly line employees have been judged by how many widgets they can produce in a day. In a manufacturing plant, everyone works out in the open, within plain view of managers. If you fail to keep up or make your quota, you're fired or demoted to pushing a broom.

Salespeople are judged by their productivity. Clerks are rated by how many files they process. CEOs are judged by the company's stock price. These are metrics that are easy to see and measure.

Unlike the assembly and clerical workers of yesteryear, today's white-collar office employee is often not under the direct supervision of a manager. He or she is increasingly located away from the office, delivering digital work product from a remote location. Other employees work in their traditional cubicles, but toiling out of sight of a supervisor, while some work a few days at home and a few days in the office. Hybrid work gives employees more freedom to conduct their business day on their own schedule, which many see as a significant benefit.

This hybrid work model creates challenges for managers, particularly when it comes to understanding their employees' productivity. They lack the insights they would gather when employees worked openly and delivered measurable product. They can't see if their employee—whether remote or on premises—is focused on the company project or watching cat videos on YouTube. He or she may even be using the company computer to send out their résumé to apply for a job elsewhere.

The advent of digital technology has given employers powerful new tools for collecting data on the daily interactions of employees with the tools they use to do their jobs. The amount and type of data that can be collected is astonishing, right down to the last keystroke or restroom break.

Snoopware Is Big Business

These new technologies fall under the moniker *snoopware*, also known as *bossware* or *spyware*. They offer employers a range of features, including keystroke logging, screenshots of workers' computers, and even access to webcams.

For example, Teramind offers a suite of digital tools that track and quantify employee behavior. According to the Miami, Florida–based company, the feature called "in-app field parsing" collects granular terminal and web app activity metrics that reveal how employees utilize and navigate individual fields and field level data. "Screen capture" allows managers to use the computer's built-in camera to see employee actions in real time or in the past. "Remote desktop control" permits the boss to take remote control over an employee's computer, or disable the keyboard and mouse during a monitored session. Managers can see desktop file activity, including creation, deletion, access, writing, and transfer operations. They can monitor employee email activity and record keyboard activity, including copy+paste commands and visible or invisible keyboard entries.

Another company, Veriato, offers its AI-powered Cerebral security platform as a tool to "detect threats from employees" by "proactively recognizing signs of risk, like changes in an employee's attitude and behavioral patterns," thereby allowing managers to take action. The Cerebral algorithm alerts the manager to an employee exhibiting signs of disengagement, agitation, or the use of profanity or selected keywords, and shows related screen shots so that the manager can determine the true nature of the incident and collect the evidence essential to taking legal action.

The West Palm Beach, Florida, company also provides the now-ubiquitous snoopware services of keystroke logging, network monitoring, file tracking, chat activity, and even psycholinguistic analysis, which identifies and categorizes opinions expressed in email texts, revealing if the employee has become "disgruntled" and a possible security risk.

Based in Austin, Texas, ActivTrak says it takes a more employee-friendly, "ethical" approach that claims to boost productivity without the use of intrusive employee-monitoring technologies like keystroke logging and video surveillance. Nevertheless, the company uses the same basic snoopware tools that have become common to monitor employees, including assessing the total time an employee spends on productive and unproductive activities, revealing when employees are actively working or taking breaks throughout the day, and analyzing productivity for individuals or teams throughout the day.

Companies like ActivTrak are careful to reassure employees and managers that Big Brother isn't looming over their shoulders. They present the surveillance as a tool for increased team productivity and a way to help team members become more productive. If used properly, such tools could be an asset to a responsible company.

In low-level jobs, monitoring has become ubiquitous. Amazon is the poster child, where the second-by-second measurements are notorious. But it's being used to oversee UPS drivers, Kroger cashiers, and millions of others. According to an examination by *The New York Times*, eight of the ten largest private U.S. employers track the productivity metrics of individual workers, many in real time.

Digital productivity monitoring has become commonplace among roles that require graduate degrees and white-collar workers. An increasing number of employees, whether working in the office or remotely, are subject to a variety of snoopware trackers, productivity scores, "idle" buttons that need to be minded, or just the drip-drip of constantly accumulating records. Gaps in one's active work record can incur penalties, from docked pay to lost jobs.

At companies such as J.P. Morgan, tracking how employees spend their time, from composing emails to making phone calls, has become routine practice. At UnitedHealth Group, low keyboard activity can affect compensation and sap bonuses. Some radiologists see scoreboards displaying their "inactivity" time and how their productivity compares to that of their colleagues. Government jobs can be tracked, too: In June 2022, New York's Metropolitan Transportation Authority told engineers and other employees they could work remotely one day a week if they agreed to full-time productivity monitoring.

Tracking can now be applied to the physiology of the employee. Microsoft holds a patent titled "Emotion Detection From Contextual Signals For Surfacing Wellness Insights." The software giant describes a "wellness insights service" that amasses biometric data from a range of wearable devices, including fitness trackers, digital assistants, and smartwatches.

Blood pressure and heart rate monitoring data obtained from wearables can assess an employee's stress levels during routine work tasks, including drafting and reading emails and attending meetings. If the employee registers elevated anxiety or stress, the wellness insight service may trigger an intervention related to the work event.

The nonlucid leader might hear about this technology and immediately think, "This is terrific! It's just what I need to get my lazy employees to work harder! No shirking at my company—no sirree! Where do I sign up?"

To that we say, "Not so fast."

Snoopware comes with two significant liabilities that employers need to appreciate.

It's Often Inaccurate

Across industries and incomes, the most urgent complaint is that snoopware is often just wrong. It misses productive offline activity, it's unreliable at assessing hard-to-quantify tasks, and prone to undermining the work itself.

Reports are common of social workers being marked "idle" for lack of keyboard activity while counseling patients in drug treatment facilities. Solving a difficult problem for a customer sometimes takes time that cannot be tracked. Working out a human resources issue may not count. And trying a new idea—an innovation—that ultimately doesn't work could be perceived as wasted time.

In "The Rise of the Worker Productivity Score," Jodi Kantor and Arya Sundaram tell the story of a finance manager named Carol Kraemer. She was not a file clerk or a call center customer service rep—the kind of lower-level employee you might assume would be subject to productivity tracking. She was a senior vice president, earning $200 an hour while she worked remotely.

When she received her first paychecks, the amounts seemed low. What was the problem? Some clerical error in human resources? No. The problem was that her new employer used extensive monitoring software on its all-remote workers. She, and others, were paid only for the minutes when the system detected what it considered to be *active work*. That meant when Carol was doing something that was visible and measurable, like typing on her computer.

In Carol's opinion, the software did not come close to capturing all of her labor. Any "offline" work, such as reading printouts, doing math problems on a calculator, or just thinking about a problem, didn't register and required approval as "manual time." In managing the organization's finances, Carol regularly interacted with a dozen people, but those interactions didn't always register with the algorithm. If she forgot to turn on her manual time tracker, she had to make a special request and explain what she was doing that warranted being paid.

She said sometimes she resorted to doing mindless busywork to accumulate clicks.

"You're supposed to be a trusted member of your team, but there was never any trust that you were working for the team," she said.

It's Demeaning

Let's face it: It's one thing to work in an open office, surrounded by colleagues who can generally see what you're doing and will know if you're sitting at your desk playing Candy Crush Saga instead of working. It's something else entirely to feel as though Big Brother were tracking your every keystroke and every trip to the restroom and will dock your pay for every measured infraction.

In general, in the United States, the laws and the courts favor the employer. While the U.S. Constitution contains no express right to privacy, the U.S. Supreme Court has historically upheld an implied right to privacy *at home*, but *not at work*. Courts favor the idea that since the company owns the equipment and the office space, it has a right to monitor its employees to prevent misuse of that equipment and space. Even if workers use their personal devices at home, their employer could still legally track their activity if they're using company email accounts, networks, or servers.

While employers are required to inform employees that they retain the right to monitor their behavior, these notices can be vague and buried in the fine print of the employee's contract or company handbook. There's no requirement for employers to tell workers specifically what monitoring programs they're using or what sort of information they are gathering.

Here's the basic rule: If the employer *owns* it, the employer can *monitor* it.

In many companies with particularly heavy-handed surveillance programs, employees are resisting. Office workers are echoing complaints that lower-paid employees have voiced for years: They say they don't have control and their jobs are relentless, without even a spare few minutes to use the restroom.

In 2019, Barclays Bank piloted a software system called Sapience, which, according to the vendor's website, gave companies "insights into work patterns" and tracked productivity by monitoring employees' computer use. The bank would even send intrusive and condescending messages to workers such as, "Not enough time in the Zone yesterday." In February of 2020, after negative staff feedback and critical media reports, the bank backtracked and announced that moving forward, the software would collect only anonymized data.

As Camilla Winlo, director of data privacy consultancy DQM GRC, said, "The problem isn't the monitoring itself, but the fact that the intrusion into employees' privacy doesn't match the scale of the threat. Although most detective controls—those used to identify risks—will require some trade-off between workers' privacy and their safety, such tools can result in monitoring not just employees' work habits, but their overall lifestyle choices . . . workers may also be legitimately using a computer in a private capacity from their own home."

Commander's Intent

This is the bottom line: Whom are you hiring? And how are you training them?

The foundation of lucid leadership in a time of chaotic change is that you create a culture of trust in your organization and you hire people who will fit into that culture. You train them to embrace trust and expect it every day. If you cannot trust someone to do their job, don't hire them.

Here's a powerful example. In the U.S. Marine Corps, trust between all Marines, regardless of rank, is paramount. This foundation of trust is expressed in a concept known as "commander's intent." This means that the individual Marines who are told to complete a mission are provided with a set of instructions, which they are empowered to modify if conditions on the ground require it. For example, a captain says to a rifle team, "Your mission is to take and hold that hill. You will attack the hill from the south." But if the rifle team gets to the target and sees the south side is impassable, while the north side presents an opportunity, the team leader has the authority to say, "We're going to attack from the north."

It's the responsibility of the leader of the rifle team to be lucid, to see the problem, and to execute the commander's intent, which is to capture and hold the hill.

In such situations, trust and lucidity are mandatory.

Distrust is inherently an expression of a false reality. By this we mean that you hire a person, you train them, and you give them responsibility. You say to them, "These are the goals we need to reach. We will give you the tools—the computer, the phone, the office—to reach those goals. You are empowered to use your best judgment. Get the job done!"

So far, so good! But then you say, "Oh, and by the way, *because we don't trust you*, we're going to monitor those tools we provided. Our computer will be watching you and tracking every move you make. Have a nice day—and be loyal to the company that loves you!"

Sounds crazy, right? But that's exactly what nonlucid companies are saying to their employees. They think that they can treat their employees like prison inmates, and then those employees will reciprocate with loyalty and hard work.

Fat chance.

To be fair, let's consider the opposing viewpoint. We can hear CEOs saying, "How are we supposed to know if our employee is goofing off while on our payroll? The employee may even be working at home, out of sight. We need these surveillance tools!"

Go back to the commander's intent. You solve the problem very simply. You say to the employee, "Your goal is to complete this report/project/chart/draft by tomorrow/next week/every month. Can you do that? Yes? Good. We will leave you alone to get it done."

You can say this to the employee regardless of their physical location. They could be working in an office you own, at home, or on a sailboat in the Caribbean. It doesn't matter. You give them the job and they do it.

The opposing viewpoint might reply, "But wait! I'm trying to be super-lucid and know everything that's going on with my employees during this time of chaotic change. The more information I have about their activities, the better!"

If your employee were a machine, we'd agree. We *need* performance data from machines. Take our automobiles. It's amazing that every new car has a full suite of computer diagnostics to alert the owner of the slightest problem. Do you have a tire that's underinflated? The diagnostics will tell you. Are you drifting out of your lane while driving? The computer will beep you a warning.

But—and we'll only say this once—*people are not machines.* Even assembly line and data entry workers, who perform machinelike tasks, are not machines. The people who pack boxes in Amazon fulfillment centers are not machines, despite Mr. Bezos's dreams to the contrary.

Human beings are driven by emotion as much as by logic. The lucid leader in a time of chaotic change requires accurate information about how his or her employees *feel* about their jobs and their lives. The happier the employee, the more dedicated and loyal they will be.

The lucid leader will take the time and effort required to get to know his or her direct reports (at the very least) and build a culture of trust.

This culture must include the most remote people on the lowest rungs of the pay scale. Remember—trusted people work harder and deliver better results!

Take Action!

✗ Advances in AI technology allow leaders to become more lucid and agile in regard to their ability to make decisions. AI algorithms can collect information about complex events, process that information, weigh the choices or solutions, and offer a recommendation— which the human leader is free to accept or reject.

✗ AI is becoming a core aspect of business strategy. The keyword is "strategy." Your organization must develop a plan for how you can leverage AI and how you intend to measure the return on investment. The AI strategy must be connected and coordinated across the enterprise, and in close alignment with the overarching business strategy.

✗ Just because it's AI doesn't mean you should embrace it! Tools are just tools, and they can either be used constructively or misused and cause damage. Be judicious with employee surveillance programs. They don't give you the full picture of the employee's activities, and they can cause a feeling of resentment. Instead, hire your people carefully and focus on developing a culture of trust.

CHAPTER 6

Stop Focusing Only on What You Think Customers Want

Does that sound crazy? After all, creating new experiences and products and services, and beautifully delivering those to customers in a way that they want, is absolutely the critical aspect of what it means to deliver exquisite customer experience. But it turns out that if we focus *only* on what customers want, as many organizations do, they're operating from the assumption that their current state of customer experience is acceptable.

The twenty-first century economy embodies hyper-competition and hyper-consumerization. Customers demand complete freedom from friction. They want price transparency. They want every transaction to be seamless. So if we assume that our current business model and the way in which we deliver value in every possible way—from the quality of our products, to our product packaging, to our distribution and supply chain methods, to everything—are all okay, and now we just need to find out what our customers want, we won't get where we need to be.

The Bar Keeps Getting Higher

The overwhelming majority of your customers receive you might call the *baseline level of current expectation*. In other words, you're giving them what you think they want. The problem is in a time of massive change the bar continues to rise, and customers are wanting far more than they used to. They want to be surprised with exquisite experiences and value. Organizations operating in that danger zone of the baseline level of current expectation are likely going to fail—maybe tomorrow, maybe next year, but ultimately, that danger zone is real. And it's what's killed many of the best organizations on the planet. So let's think about why we would ask ourselves the question, "What do our customers hate?" as being just as important as, "What do our customers love?" And that's the thesis of this program: Much information can be gleaned when you look at the flip side of the coin. In other words, if you find out what a customer hates, then they've just told you what they want in a clear and crisp way.

Everyone Looks for the One-Star Ratings

A powerful reason why it's extremely important to think about the haters as much as the lovers is that haters can cause revenue and customer defection. In a time of digital ubiquity, and what Google refers to as "micro mobile moments," we use our connected devices to make decisions about where we go to dinner, what hotels we stay in, and what products we're going to buy. And one of the ways in which we make those decisions is what we call *hyper-influential social communities*. For an example, on Amazon, when you're thinking about buying a product, the first thing you're likely to do is find out how other consumers rated the product. The larger the crowd, the more authentic the rating, so you take a look at the ratings. Now, here's a really simple question: Do you skip all of the five-star ratings and go straight to the one-star rating? Most of us do that because we want to know what the *haters* think. If something is wrong with the product, we'll find out by reading the bad reviews. The haters are some of the most influential customers you have. Isn't that interesting? The people who hate what you do have the biggest influence over your success.

The solution is to stop making customers hate you. Avoid getting one-star reviews.

How do you do that? You begin by finding out what your customers *hate*. Seems pretty simple! But virtually every discussion on the topic of customer experience has to do with trying to find out what customers *want*. And the truth of the matter is customers really don't know what they want. And you know what? It's not their job to invent a better experience. That's *your* job. That's where the heavy lifting comes in. Surveys and promoter scores and other tools are fun. They create great graphs and charts. They're easy to use. They're supposed to be best practice, but really understanding the haters is the secret to developing the best customer experiences on the planet.

Haters Will Deflect Your Customers

The pop star Taylor Swift famously said in one of her top songs, "the haters are gonna hate." But here's the bigger problem with haters: They're *deflectors*.

What does that mean? They will destroy your business if you don't identify them and fix the hate.

There's something called the "bumper sticker syndrome." Have you ever noticed that the people with the worst ideas have the most bumper stickers? Well, it turns out that haters are very prolific, and they are loud. Haters have big voices. They're part of the "loud crowd." And the loud crowd can't wait to talk about how much they hate you.

In our world, there are plenty of trolls. These are negative, destructive people who just want to say something bad about somebody else. But generally speaking, the data that comes in and the rating aggregation of the trolls are usually not statistically meaningful. Every business has its share of trolls.

Your genuine, sincere haters have the power to deflect business from you. If your social ratings are low because customers hate you, and you have not resolved those hate points, then you will ultimately fail. You cannot afford what we now call digital deflection. But it's not just there you get deflected; you can be deflected anywhere along the customer touchpoints. And understanding what customers hate is far more insightful than basking in the praise customers heap upon you.

The moral of the story here is pretty straightforward. If you really want to know what customers want, you have to first of all look at the flip side of the coin, which is what they hate. That's where they tend to speak far more accurately.

When it comes to what your customers really want, that's your responsibility to uncover. It never showed up on a survey at Apple computer that Steve Jobs should have invented the iPhone. That insight came from a keen and unique and special understanding about the company's users. In fact, even today, when you look at the Apple Store, they have applied the simplicity of their graphic user interface to the way in which people experience their retail environment. As a result, the Apple Store is one of the most profitable retailers in the world. They're responsible for knowing what their customers want, and they know they're not going to get there with surveys.

Surprisingly, you see former executives from some of the worst corporations ever to exist in America, who are no longer with their companies, going out and teaching people how to deliver exquisite customer experience. You see failing hotel chains teaching customer experience programs, and amusement parks that have lost their way teaching courses on customer experience. It's insane! And unfortunately, many leaders, executives, and well-meaning companies don't honestly know where to turn. Because there's an agenda: Everything's going to be just fine. If you buy our software, if you buy our training package, if you let our consultants climb around your business for a few months, everything's going to be fine.

"What Do You Hate About Me?"

But customer experience is holistic. It begins with having an amazing work environment based on a culture of happiness. It's about really being honest with ourselves and asking ourselves some tough questions. And the toughest question is, "What do you hate about me?"

When was the last time you uttered those words? Probably never. Because you probably do everything you can to make sure nobody would ever use such a strong word against you. But the best organizations on the planet accept the fact that they're imperfect. They know that they do not serve a so-called customer, some monolithic archetype. They know they serve a wide range of personas. And they don't look at those personas from a perspective of demographics. They look at those personas from the perspective of what each person hates and loves. They also look at those hates and loves across the five touchpoints. And of course, they look at them from the perspective of digital and physical environments. It's really hard to ask the question, "What do you hate about me?" but it is the most powerful thing you and your organization can do to rapidly scale exquisite customer experiences.

The Eternal Question: Do Customers Always Know What They Want?

In a perfect marketplace, the customer would always know exactly what they wanted, and the business would be able to provide it. The two—customer and business—would move in tandem, seamlessly, like Fred Astaire and Ginger Rogers doing a dance number in a Hollywood film. Each would keep step with the other, with no friction. Whatever the customer wanted, the customer got; and to reduce waste to zero, the business would never produce anything the customer didn't want. The business would produce exactly what the market wanted—nothing less and nothing more.

That would be pretty good!

But then to make the customer experience even better, imagine that the business knew what the customer wanted *before the customer did*. Wow! Imagine that!

The customer would say, "Gee, I want something . . . not sure what it is . . . something to make my life better . . . but I cannot describe it."

The business owner would reply, "Ah! I have exactly what you want. Here's the Gizmo 100! It's available right now!"

The delighted customer would exclaim, "OMG! The Gizmo 100 is exactly what I wanted but could not describe! Thank you!"

If such a system existed, it would be a win-win for both sides. The customer would be continually amazed and delighted, and the business could charge a premium price for its innovative products.

In fact, there are many businesses that try to do exactly that. Not only do they strive to produce what the customer wants, but they try to *anticipate* customer demand by producing goods that their customers either cannot articulate or would dismiss as being impossible.

The notion of the visionary leader who can conjure out of thin air products that people never knew they wanted until they saw them is attractive and powerful. It's also really *easy!* Why sweat over market research when you have a crystal ball?

Reality is far more nuanced. Customers are unpredictable, and the best you can do is to be a chaotic changemaker and be agile enough to zig when your market zigs and zag when it zags.

Take Action!

✗ The overwhelming cause of customer experience and ultimately business failure was the lack of a formal customer experience strategy coupled with a comprehensive human experience strategy that addresses the experience for employees and your total experiential ecosystem. Simply stated: Failing to plan is planning to fail.

✗ Go beyond outdated surveys that provide nonactionable, fractional insights, and collaborate with customer-facing employees to identify the real opportunities.

✗ Three things are necessary to win at customer experience, and the overwhelming majority of organizations typically are only good at one or two of these three vital practices. This includes the following:

1. Get keen insights about what customers love and hate that go beyond outdated surveys.

2. Develop a formal customer experience strategy, and then act upon it daily.

3. Provide job-specific customer experience training across the three categories of employees to include leaders, managers, and staff.

PART II

Innovation

After leadership, the next most important weapon in your arsenal to stay ahead of chaotic change is innovation.

Innovation is defined as:

The creation of new value that serves your organization's mission and customer

Innovation begins with the spark of *any* new idea or process from *anywhere*—the scientist in the laboratory, the marketing manager, the techie in IT, the assembly line worker, the maid who cleans the hotel rooms. This new idea or process—large or small, planned or a lucky accident—is then evaluated against a simple standard: Will it add value, help us fulfill our mission, and serve our customer?

But it cannot stop there. Today's amazing innovation is tomorrow's status quo. The product, system, or service that is disruptive today will soon become vulnerable to disruption by something newer and more powerful. This is why in the era of chaotic change, innovation must be a sustained, never-ending, and ideally systematized process.

Let's dive into the three key phases to creating a practical and durable system of innovation in your organization.

The first is your *innovation mission*. This is the overall road map that will guide your efforts. It's not unlike the mission of the organization as a whole; the difference is that it's focused only on innovation. A big chunk of this section will be devoted to your innovation mission, because it's like the foundation of a skyscraper: It's got to be rock solid.

Your innovation mission provides the direction for your *innovation operating system*. Just like the operating system in your computer, it manages all the moving parts that go into a robust system for producing and exploiting new ideas, inventions, and processes.

The heart of the innovation operating system is your *innovation pipeline*. It's the step-by-step process whereby the sparks of new ideas become bright shining stars of innovation. The idea of a pipeline should be familiar to you, as no doubt you've already got a sales pipeline for converting prospects into customers, or a hiring pipeline for screening job applicants, evaluating them, and eventually onboarding them. Your innovation pipeline is no different and should be a ubiquitous part of your everyday operations.

CHAPTER 7

Create a Culture
of Innovation

Culture is how an organization thinks, feels, and works as a whole to achieve excellence. It's about pride, passion, and environment. It's about whether the company provides a functional environment for employees to succeed and thus for innovation to succeed. It's about having the right empathy and focus on customers that allow their input and needs to get into the company brains in the first place; an organization that doesn't listen to customers won't make the right customer-focused decisions and will thus be doomed to fail.

Love requires the same essential ingredients: a functional environment; a passion to think, feel, and work together; a consistent empathy; and, above all, the right focus. So consider this a chapter about love; only in this case, it is a love of customers that translates into a loving organizational culture that allows other great things to happen.

External Focus

The starting place for both love and culture is focus. Without the right focus, all other efforts will be misdirected or misguided, and the best intentions will miss the mark or go unnoticed. As a result, the organization must start out focused in the right place: external focus.

External focus for an individual is simply focus on the other, not on oneself, and focus on listening with unfettered attention to the other person, with empathy, compassion, validation, and a top-to-bottom satisfaction of that other person's needs. In an organization, external focus is an unfettered focus on the customer, which also entails listening, empathy, and especially a top-to-bottom satisfaction of the customer's needs.

Beyond satisfying these needs, in culture as in love, delivering above expectations brings delight and substantial reward to both parties involved. In the case of both individuals and organizations, love and culture bring about the right decisions—in the former case personal and the latter business decisions—to bring mutual reward and beyond that, the visceral connections and loyalty that if managed properly can last forever.

You may ask: "How do we, as an organization, get external focus?"

It's a great question. If you give a PowerPoint presentation in a conference room about external focus, or wander up and down the organizational halls whispering "external focus" in the ears of the occupant of each and every cube, will external focus happen? Maybe, for a while. But will it be sustained? Will there be a permanent commitment to external focus? Will it be ingrained and behavioral? Probably only until the next set of forms arrives that accomplish some mundane organizational task, or worse, that control the risk of the project or process you're working on.

Further, it turns out that almost everything wrong with innovation in an organization has, at its roots, *internal focus* as a cause. Excessive risk management, delayed or slow launches (or no launches at all), inadequate resources, solutions that don't meet or exceed customer needs and thus fail in the marketplace—all can be attributed to internal focus.

The solution to creating external focus is simply to eliminate internal focus. Leadership that avoids internal focus creates external focus.

Employees are more likely to become devotees and enthusiasts, which self-perpetuates into more external focus, and you have an oxygen-giving positive spiral of good feeling and external focus that satisfies customer and organizational needs alike.

Again, the solution is pretty simple but bears repeating: To achieve external focus, get rid of internal focus.

The Innovation Focus Model

In this model, there are two wheels, one of internal, one of external focus. The internal focus wheel features such components as process focus, risk focus, blame focus, tech-centered focus, and a reactive mindset. None of these features of internal focus should be of any surprise to readers who have followed along. And the negative effects of these features on innovation, similarly, do not need to be explained again.

On the external focus wheel, one sees results instead of process, opportunity instead of risk, accountability instead of blame, customer-centered instead of tech-centered, and most of all, proactive instead of reactive. As a manager and leader, you must try to keep your organization on the right side of this chart.

Innovation "Socialists" and "Capitalists"

At the top of the chart there are the terms "innovation socialists" and "innovation capitalists," which correspond to internal and external focus, respectively. In fact, many organizations do not reward their innovators for the successful commercialization of their technology; the technology becomes an end in itself. Activity, not results. Many of these innovators aren't punished—or even measured—for their lack of results. The end result is often—nothing.

Internal focus can lead to this end. If you start and kill ten innovations because of adversity to risk, then the internally focused organization still feels like it accomplished something, because it did start the innovations, and because it did go through the motion of killing them. But these false starts do nothing for the customers; they do nothing for the external world. While the process was executed and it worked, it ultimately sucked resources away from things the outside world, and thus the organization, really needed.

I refer to external focus as innovation capitalism. In innovation capitalism, innovation must start, grow, and launch based upon its own merits. That means it must be externally focused and externally driven to the point of generating the return on investment and income necessary to sustain it. To do that, in a capitalist society, it must offer something better than the competition and better than before. To do that, coming full circle, it must be sufficiently focused on the outside world to know what better than the competition and better than before really is.

It's so simple: A culture existing on the right side of the model is aligned for success; one on the left side will eventually collapse under its own weight.

Success Referencing

Companies with strong innovation cultures also tend to success reference rather than failure reference. You can see it in the language the company uses in meetings. You can even see it in the body language. Success referencing occurs when a company talks about its successes and measures new products or ideas against its successes. A new idea is assumed successful until proven otherwise, and there's an almost childishly excited hope that the new idea will have the same success as the old one. The idea gets a champion, gets sponsorship, and gets resources almost right off the bat. People buzz about what the new product and its adjuncts could really become.

Failure referencing, on the other hand, occurs when a company always looks back on failures and uses those as a model for what might happen to a new idea. "We can't do this because . . ." It becomes the mind-set in which everyone in the room plays "find the flaw" with the idea, because that's the way they can look smart and avoid personal risk at the same time. It's also easier. It is easier to "failure reference" than to "success reference."

If a company holds up its own innovations with pride and tries to fit new ideas into something resembling a success, the culture is right. If a company lives constantly in a world of fear, risk, and doubt about a new idea, especially based on some past experience, the innovation culture needs some work.

To develop a robust culture of innovation, think positive and reference success.

Achieving Collective Passion

In assessing collective passion, the first question is whether or not a company really embraces or internalizes the importance of successful innovation as part of its long-term success. If it doesn't—that's big trouble. The answer, of course, is a long-term view of success and a thorough understanding of the company's customers and what value they might want over the long term. To get to these places, a company must adopt—you guessed it—an external focus.

Aside from that, passion is really all about whether the company and its employees truly have passion for their customers and their products. Do they understand the customers? Empathize with them? Feel good when the customer feels good and feel bad when a customer feels bad? Really good, smart people with balanced personalities and good training do these things. These concepts are hard to teach, but again, it's about putting oneself in other people's shoes. It's about external focus.

To get there, it's important to hire the right kind of people in the first place—positive, enthusiastic, energetic, and so forth. But it is also important to stress the importance of the customer, and provide a reward and internal social system that fosters these feelings. Let people talk openly about customer successes and failures. Allow them the time to find out for themselves whether customers are having good or bad experiences. Send them on innovation safaris. Make each employee feel like every customer is both their friend and their responsibility. Reward them for doing so.

And finally, do away with the dry, dull, corporate-speak in visions, mission statements, and strategic and tactical plans. Make them fun and active.

It's hard to instill passion in a person, but if you create the right environment for passion, whether with corporate culture or personal love, great things can happen.

Achieving Craftsmanship

Craftsmanship is like passion, but it is more specifically directed at the product, not so much at the customer. The idea is to produce products that not only meet but exceed customer expectations or, more colloquially, make the customer say "wow."

How you get there, of course, starts with external focus. A company or an individual within the company will craft the best products or customer experiences when they can feel that experience personally. When a customer says "thank you" or gives pleasant compliments about the product or service, you're there; when they come back (loyalty) or refer others (evangelism), it's even better.

A culture of craftsmanship comes about when management has a get-it-right mentality—and gives employees the time, resources, and recognition to do it. When employees are treated like entrepreneurs, and they feel that the company is their own personal small business, they will naturally deliver a higher standard of quality and take more responsibility for their products. So giving employees the time and empowering them to get things right goes a long way toward craftsmanship.

One of the best ways to instill the idea of craftsmanship and entrepreneurship in an organization is to give employees an ownership share or a share of the profits. In large companies, of course, this can mean shares of stock or options or profit-sharing bonuses. In smaller companies it can also mean bonuses and shares of profits, and it may also be ownership depending on how the business is set up. Either way, when employees have skin in the game, better results almost always follow, especially in the area of craftsmanship but also in the area of passion just described.

Achieving "Fear No More"

You must create a culture where employees can take smart risks. They can start—and end—projects when it makes sense. It's a culture where failure is rewarded and sometimes even celebrated as a necessary step to success. Psychologists sometimes group people into two types: people who live in fear and dispense fear all their lives, and people who live in love and dispense love all their lives. We all know the types. And to get the right culture, management needs to do the latter (that love thing again). Positive reinforcement, and even delivering bad news with a soft touch and a "here's how things can work better next time" can help. Managing fear properly in an organization is a lot like providing a positive environment for children, where they can make mistakes, learn from them, and not feel like they have to lie or work around you to get what they want.

Also—recall the discussion of the benefits of failure. A healthy organizational culture recognizes the new health that failure—tried, recognized, and learned from—can bring to an organization. Such an organization has a smart-risk culture. Remember, a tree pruned of its dead branches comes back healthier for the experience. Have a failure? Take the team out for drinks, talk about it, and celebrate it.

Getting the Team Right

Some businesses may be small enough to carry on as one-man bands, but it's not likely to be the case for long, especially if a business is destined for success.

Successful innovation is highly dependent on having a good team, and the right team, in place, with the right mind-set. The right team has energy, collective enthusiasm, and diverse skills and personalities. They are interesting people, and they interest each other. And they can all put their own interests and rewards aside to honor one common bond: the interests of the customer. They are externally focused individuals, other-centric individuals who interact easily with others and realize that in many situations, one-plus-one equals three.

Selecting such people isn't easy because these attributes are almost indiscernible on a résumé. It's all about finding out what kinds of experiences these people have had with others, and with helping others—whether in a business or rock climbing on a 1,000-foot granite face. Signs of customer involvement and team involvement in the past should be observed. And of course, positive, forward verbal language (*we*, *us* instead of *I*, *me*, *they*, and *them*) and body language are all pluses.

A successful team culture shouldn't rely 100 percent on the individuals in the team. Of course, the environment created by management is extremely important. People need the chance to act and interact in positive team settings—at work and at play. Team-oriented organizations have group functions, basketball courts, gyms, and organized after-work events. Teamwork is encouraged—and teams are built through team-building events. Team-oriented organizations consider these items as investments, not expenses, and they are all part of the investment an organization makes in keeping ahead of the game.

The Nine Characteristics
of Effective Cultural Leadership

It wouldn't be surprising if, by now, you've concluded that culture is really a matter of leadership. External focus, passion, craftsmanship, creating a safe environment in which to innovate, and building the right team are all outcomes of smart, properly focused, compassionate, and effective leadership. Specifically, here are the following nine characteristics of effective leaders and the leadership style they deploy in the innovation space.

1. Keep and Share the Focus

We've made much of the external focus that gives oxygen and direction to innovations. Leaders must get focused, stay focused, and show that focus to their teams.

2. Keep a Long-Term Perspective

This one's easy: A leader focused exclusively on short-term results will get just that—and a company that runs out of ideas fails in the marketplace longer term. Keep your eyes on the prize, which is almost always long term.

3. Understand the Customer

Sounds simple, but how many leaders really understand their customers? All of their customers, not just the biggest ones? How many have taken the inverted pyramid to heart? How many take the time to really get into the heads and into the experiences of their customers? How many of them innovate by walking around? Do top management presentations or company annual reports reflect ideas taken straight from a customer's experience? Well, there are a few such leaders, but most rely on cumbersome market intelligence, or worse, try to dictate what the customer needs. These processes are seldom fast enough; they miss market opportunities—or worse, tick customers off.

4. Understand Innovation and Innovators

Leaders are simply more effective when they understand who and what they're leading. Do leaders in your organization really understand how innovation works? Or do they let their "innovation program leads," where they exist at all, do that for them? Leaders who have "been there, done that," or leaders who take the time to work in the kitchen, so to speak, will gain a better understanding of what's eventually served up.

5. Passion, Patience, and Perseverance

An assortment of leadership qualities are combined into a single bullet—for a reason. Passion without perseverance doesn't work, and without at least some patience, the organization will go into stop-start mode and overreact to every new input. Leadership is really about a combination of good elements and traits without too much emphasis on one trait; otherwise, it looks false or contrived.

6. Always Think Win-Win

Good leaders look for scenarios where everyone can win—the customer, the organization, the employees, the shareholders, the leaders themselves. Out the window go such examples as Merrill Lynch's John Thain, who had the gall to complain about not receiving a year-end bonus from his firm, which had lost billions. His argument that he prevented more severe losses rings hollow when considering the "win-win" principle. And in 2008, former GM CEO Rick Wagoner—as justified as he might have felt in taking the corporate jet to Washington, hat in hand, to ask for a bailout—clearly sent the wrong leadership message to everyone inside and out.

7. Be Quick and Patient

These seemingly opposing traits actually do work well when worked together in the right way. At the same time, impatience overdone is a problem—that is, leaders who don't appreciate that good strategy may take some time to implement and that initiatives need room to develop and mature. Such leaders will create frustration and stress in those beneath them.

8. Be Truthful

People—both customers and employees—are smarter than you think. They want honesty and can generally see through the alternatives. Employees will respond better when a CEO admits a mistake or a hard truth about their organization or its innovations. Jeff Bezos has often been quoted as saying one of the key elements of being a good business leader is the capacity to tell hard truths, not run away from reality. And we all know what troubles can be caused when a company tries to hide its failures from its customers, as the recent Volkswagen emissions debacle clearly illustrated.

9. Have Fun

There is often too much seriousness and dryness in corporate environments. The best work is fun work; that is, when you're having fun, you put more energy and thought into your work. You may not have always observed it, for when the straight-laced, dark-suited bunch get together for an off-site, all forms of productivity melt away into silliness—but it's because these people don't have much fun during normal work life, so they overcompensate. Fun leadership begets happy employees and expansive thinking.

It may seem odd to incorporate a section on digital innovation in a chapter on culture. But not only has the world as a whole gone digital, but so has innovation. As in other aspects of business—marketing, operations, customer service—digital media will open the doors to effective innovation, and close doors to those who choose to lag or ignore it altogether.

Today there is a tremendous opportunity to speed customer input and innovation efficiency by using digital media as your platform. The idea of a digital command center is to be able to acquire great information quickly to aid in the innovation process, and to link innovation and other marketing and image activities in such a way as to improve your company's reputation and even to manage the brand.

A holistic digital innovation model incorporates several digital platforms in and around innovation.

Key Components of Digital Innovation

Listening Posts

Listening posts were introduced as a bonding time method to capture the buzz in the marketplace and media about your products and product concepts. Listening posts are easy to set up as keyword selectors for various media. For example, you can set up a listening post by entering a keyword—your product or your brand—on Yahoo! News or Google Alerts. It's a great way to quickly, and in real time, identify what people think about product concepts, service concepts, brands, and so on. You can also set up posts and questions on social networking sites and other new media to capture buzz and get more specific answers to questions you might pose. Listening posts are a great way to get real-time feedback—for free.

Crowdsourcing

Crowdsourcing is the process of sending out questions, or problems, to problem solvers, and providing rewards for people who deliver solutions to specific problems. Crowdsourcing can be done easily and cheaply on innovation portals like Brightidea's WebStorm. You can also manage the size of the crowd to get the kind of relevant feedback you want with micro-crowds.

Wikipedia, which has been developed with almost no centralized control, is probably the best example. It has thousands of volunteer contributors from around the world that produce articles of remarkably high quality—and the contributions are from people who aren't getting paid.

Threadless—a T-shirt company—asks site visitors to submit to a weekly design contest and then vote for favorites. The entries receiving the most votes get sent into production, and designers get royalties and prizes. In doing this, the company utilizes the collective intelligence of more than 500,000 people to design and then select its T-shirts.

There's much more to this ideal of crowd intelligence than just a "fuzzy collection of cool ideas." To unlock the true potential, managers need a deeper understanding of how these systems work and what motivates contributors to contribute. The MIT Center for Collective Intelligence gathered almost 250 examples of web-enabled collective intelligence. They looked at who participated in the organization and among the crowd, and the motivation behind the participation. Sometimes it's money, but it can also be love or enjoyment, glory or a competitive drive. Money talks—for example, InnoCentive is a company that offers cash rewards, typically totaling in the five or even six figures, to researchers anywhere in the world who can solve challenging scientific problems such as how to synthesize a particular chemical compound.

They also looked at the "what and how" behind the average project. This is the same as the project's mission or goal. It may be to create something new; it may be to decide or answer a question. Companies like Threadless both create and decide. The company found there is a difference between situations in which contributions are received or evaluated independently or in which there is a dependency between the contributions. Collaboration, which occurs when participants are working together to create something and must depend on one another to some extent, can be motivating. Wikipedia articles are a good example of independence but also collaboration—articles are created independently, but additions or editorial changes made within an article are strongly interdependent.

It was found that the crowd mentality was most useful in situations where "the resources and skills needed to perform an activity are distributed widely or reside in places that are not known in advance."

It was also found that for this to work successfully, you must be able to divide an activity into pieces that can be performed by different members of the crowd. There should also be safety mechanisms in place that prevent people from sabotaging the system. In many cases the final decision is left to a specific internal group in charge of the crowd's task, even if some of the intermediate decisions are made by the crowd.

Buzz-Building

Digital innovation also includes buzz-building. Now, historically, if you were a national company, building buzz historically could be an extremely expensive task. If you were a local company, it was slightly easier, because you had a local market and presence, and it was easier to go viral in such a market.

But now with the internet you can build buzzes and get your products and services noticed by crafting intelligent blogs, posting on social networking sites, and the list goes on. In fact, most marketing departments and marketing agencies have created a digital media section, and in some companies, the marketing department has transitioned into a digital media department altogether. It's that important. So buzz-building through social networking and other digital marketing modalities is a powerful way to sell and to be successful in any market.

That buzz, of course, can generate feedback useful to the innovation process, and more generally, to the development and strengthening of your innovation culture.

Innovation Floating

The concept of *innovation floating* is also very powerful. Companies routinely produce special and powerful websites and photo-realistic imagery on technologies that didn't even exist to send images out to the digital universe. Such images can also be presented as film clips for YouTube or other media. Floating can occur not just for the consumer space but also for the distribution and channel space. The purpose is to get feedback on the product and learn whether people would buy it if it was in the marketplace.

Prior to engaging in innovation floating, particularly with people outside of your company, it is highly recommended that you consult with a registered patent attorney to ensure that your innovation floating program does not invite problems down the road. According to patent attorney Robert Siminski, one major risk of poorly structured innovation floating programs involves "public disclosure" of an invention, which can serve as the basis for invalidating any resulting patents. Another risk is that if not properly controlled, "feedback" can find its way into patent applications, thereby calling into question who should own any resulting patents.

The bottom line is this: Properly structured innovation floating programs can provide significant value, while poorly structured programs can have serious ramifications.

Online Brand Reputation Management

Another important area that can be addressed digitally with ultimate benefits to the innovation process is brand reputation management. Reputation management is the observance of online feedback from your customers about your products and services—and the corrective action you might take to turn the buzz positive. It's extremely important—whether you're a corner liquor store or a multinational company, your reputation online is your reputation. Naturally your brand reputation, in turn, serves to nourish your innovation culture.

In fact, many in the corporate world today keep track of online feedback just like a credit score; when there's a "ding" against it, you act. New firms have been set up to specialize in online reputation management. Sometimes the information is erroneous or laden with bad assumptions, but it's still important to keep track. Online brand reputation management provides a significant proactive opportunity to really build a brand, by creating your own internet footprint that speaks favorably about your company and its products and services. But also, don't forget what people are saying, feeling, and experiencing becomes input to your innovation process.

Automated Innovation Portals

Innovation portals that capture both external and internal ideas and feedback are essential for soliciting, acquiring, managing, and reviewing ideas on the open innovation front. The key is the automation, through the filtering and toggle techniques.

To reinforce the idea, remember that to make open innovation work through such portals, you need to clearly state your needs (remember the innovation platform?). Developing a platform not only gives the necessary structure to filter out unwanted ideas, but it also helps you as an organization to focus externally and to get your culture aligned to a set of products and product characteristics that really meet the needs of your customers. A well-defined wish list is a sign of a company with a well-developed innovation culture.

So in summary, digital innovation is centric to almost all businesses today. You should be using listening posts, micro-crowd sources, buzz-building, innovation floating, brand reputation management, and most important, automated innovation portals.

Take Action!

✗ Above all else, external focus is the cultural lifeblood that keeps innovation flowing in the right direction. Without external focus, the remaining cultural elements of good innovation will be difficult if not impossible to achieve.

✗ Almost all things that go wrong with innovation can be attributed to internal focus. Therefore, creating the right culture for innovation really starts with getting rid of the manifestations of internal focus—excessive risk aversion and process focus, bureaucracy, blaming, reactivity, failure referencing.

✗ Beyond external and customer focus, effective innovation cultures exude passion, craftsmanship, safety for risk takers, and good teamwork.

✗ Today's effective innovation culture must also be digital in order to keep up with the pace of change and to properly augment customer knowledge and ultimately the innovation process itself.

✗ The best organizations in the world are comprised of employees who truly believe in the organization's mission. They also feel that the organization is part of their own evolutionary journey, and they see daily the impact they're having on others.

CHAPTER 8

Happiness as a Strategy

Failing organizations assume that customer experience is
something that's just simply robotically transacted by their organizational
team, without regard to their culture. It turns out that there's more to the
story. The best organizations have actually institutionalized happiness in
the form of Happiness as a Strategy (HaaS)—which I discuss in my other
books more extensively—and it actually works.

But it doesn't just work in terms of delivering incredible experiences
to your customer. It also allows you to attract and keep the best talent,
and significantly increase productivity, presenteeism, and employee yield.
In short, the only way you can deliver exceptional customer experiences is
to make *happiness an institutional priority*. When we love, respect, engage,
and collaborate with and support our teams, we create the life support
system that makes the customer experience real.

A culture of organizational happiness drives a culture of happy customers.

You've probably experienced a business where you can just feel the vibe
of negativity. The culture is sick and the people are unhappy, and they can't
wait to share the pain with their latest victim, also known as the customer.
This is one of the big reasons why customer experience is failing within the
majority of organizations in the world.

In a time of disruption and innovation, great organizations are reimagining the way in which they can use customer experience innovation to drive a happy organization that delivers exquisite customer experiences and quality of work life.

The High Cost of a Toxic Culture

Let's cut to the bottom line: A healthy, positive company culture is a more profitable culture. This means that even if you're a miserly Scrooge who cares nothing about human feelings, you should at least *pretend* to care, because negativity sucks the profits from your balance sheet.

One cost is employee turnover. According to a report by SHRM on workplace culture, one in five Americans had left a job in the previous five years due to bad company culture. The cost of that turnover was an estimated $223 billion.

"Billions of wasted dollars," said SHRM president and CEO Johnny C. Taylor, Jr., SHRM-SCP. "Millions of miserable people. It's not a warzone—it's the state of the American workplace. Toxicity itself isn't new. But now that we know the high costs and how managers can make workplaces better, there's no excuse for inaction."

The report, *The High Cost of a Toxic Workplace Culture*, found internal negativity created significant costs to companies in turnover and absenteeism; revealed indicators of toxic workplace cultures, such as harassment and discrimination; and highlighted the alarming impacts on employees. It found that 26 percent of employees surveyed say they *dreaded* going into work.

Think about that. What if you had 100 employees, and nearly one-third of them said they dreaded walking in the door every morning? How could those miserable people deliver amazing customer experiences? It can't happen. The only result will be a "hatepoint farm," where the only crop is hatepoints sprouting like weeds everywhere you look.

You can look at company culture as a form of capital that can be either built up or squandered. In *Harvard Business Review*, Kevin Stiroh, executive vice president and head of supervision at the Federal Reserve

Bank of New York, wrote that a low level "cultural capital" increases the possibility of employee misconduct, and must be seen as a form of risk just like liquidity risk or operational risk.

A company's cultural capital impacts what a firm produces, how it operates, and how it treats its customers. Cultural capital is just as important as physical capital (equipment, buildings, and property) and human capital (the accumulated knowledge and skills of workers, or reputational capital, like franchise value or brand recognition).

Investments in cultural capital is how you reduce that risk. He wrote, "As with other forms of tangible and intangible capital, a firm must invest in cultural capital or it will deteriorate over time and adversely impact the firm's productive capacity."

Toxicity Destroys Customer Experience

Customer experience requires that you eliminate what your customers hate so you can be the best option for potential and existing customers.

Likewise, the culture of happiness requires that you eliminate what employees hate so that you can be the best option for them. When you become their best option by respecting and honoring them, they will, in turn, serve your strategic initiatives and your customers.

Put more bluntly, there is no way that you can deliver exquisite experiences to your customers until you're committed to doing the same for your employees. You need to create a positive company culture so that you'll (1) have employees who will happily deliver outstanding experiences to your customers, and (2) be able to attract and retain the very best employees—those who want to work only for an organization with a mission and culture they believe in.

Unhappy Culture = Sick Employees

Unhappy cultures create for stressful work environments and stressful work environments result in an increase of healthcare costs by as much as 50 percent.

Research has shown a link between employee health and job performance, including workers in customer-facing roles. A growing body of evidence documents the relationship between a stressful work environment and a range of chronic conditions, including depression, hypertension, and sleeping problems.

A 2019 study by Colonial Life found that more than 20 percent of workers spend more than five hours on the clock each week thinking about their stressors and worries. An additional 50 percent of workers reported losing between one and five hours of work to stress each week.

Citing statistics from the U.S. Bureau of Labor Statistics, Colonial Life concluded that with 128.5 million full-time employees earning on average $21 per hour, workers who are disengaged or unproductive because of stress are costing employers billions of dollars.

Need More Evidence of How a Bad Culture Breeds Hatepoints? OK . . .

In case you're not yet convinced, here are some other nasty ways a toxic culture can drag down your organization:

Kills your brand reputation as an employer. Unhappy cultures repel great talent. Glassdoor ratings and employee word-of-mouth are central to acquiring exceptional talent. Exceptional talent is required for exceptional customer experience.

Organizations with a negative culture can enter into a vicious downward spiral. Bad cultures of unhappiness *attract more bad people.* Bad people deliver bad experiences, both internally and externally. Customers respond poorly to the experience, which helps feed the negativity that is the culture that attracts more bad people, and so it goes.

Kills innovation. Unhappy cultures stifle innovation due to the simple fact that stakeholders refuse to collaborate with both customers and leadership. Without this collaboration, your innovation pipeline will dry up.

Makes inclusion and diversity impossible. The entire premise of inclusion and diversity is the ability for an organization to treat all of their employees equitably and with respect. I've never seen an example of an unhappy culture that has truly put diversity and inclusion first.

How to Get Happy

Happiness as a strategy (HaaS) requires a thoughtful and measured approach. Here is a framework for building a strong and effective happiness strategy within your organization.

Happiness mission statement. While your "normal" mission statement is aimed outward, at your market, your happiness mission statement needs to be aimed inward, toward your organization and its stakeholders. Taking good care of your employees is just as important as taking care of your customers. After all, your employees are the ones who create the value that attracts, delights, and keeps your customers. You need a mission statement describing the culture you want to create.

Specific happiness goals. Yes, happiness can be measured. Perhaps not directly, but in its effects on employee behavior. Some key employee happiness metrics include the employee turnover rate, absenteeism, productivity, mistakes made, and customer complaints. Find out what they are today, and set goals to improve them by set points in the future. Remember, you cannot "buy" employee happiness with cheap gimmicks like handing out gift cards to Starbucks. Your company culture must be genuine and deeply rooted in your everyday interactions.

Measurements beyond surveys. Employee surveys are just about as useful as most customer surveys; that is, they are useless. According to a study by Officevibe, 70 percent of employees don't respond to annual engagement surveys, and nearly 30 percent think the surveys are a waste of time. To make it worse, 80 percent of employees don't believe managers will act on survey data.

Companies should look at using "micropulses" (a few questions focused on a particular topic) and key touchpoints to sense how employees are feeling. In this way, companies can maximize insights while simultaneously minimizing the use of employee time.

Continuous feedback means continuous *listening* and should translate to continuous action. Employee happiness levels can be gauged by listening to employees' signals, using the subsequent insights to take action, and then communicating back to employees that action has been taken. For employees, a particular source of unhappiness is their belief that management will take a new idea submitted by an employee and then appropriate it as their own. That's a sure way to generate a truckload of hatepoints!

Find out what your employees hate. Yes, this can be painful, especially if you—the leader—are identified as being a source of hatepoints. It takes a thick skin to log onto Glassdoor and read what some of your employees have to say. But it's always better to bring issues out into the open.

You need to consider that your employees may be reluctant to disagree with you out of fear of retribution. Many companies have a forced "happy" culture that boasts of "open communication" as a corporate value while managers discourage dissenting opinions. Open communication is built on a foundation of *trust*. First comes trust, then comes truth.

Eliminate what your employees hate. This is the moment of truth when your employees will say, "Just as we thought! We make suggestions, and the boss ignores them." Or they will say, "Wow! We made a suggestion, and the boss listened!" Ideas are useless unless followed by action. If you're not going to respond to your employees, then don't even ask them. But you should ask them, and you should respond, even if to say only, "Thank you, but we can't do what you ask. But we appreciate the suggestion."

Find out what your employees love. This is the flip side to finding out what they hate. But it's important that you don't waste your time and money on initiatives that your employees may not care about.

For example, as Nishal Mistri from Curious Thing noted, an increasing number of employers are turning to cheap perks like free food, beer fridges, Ping-Pong tables, and bean bag chairs in offices, with the idea that these things will deliver a positive workplace experience, particularly for younger employees.

Fully 40 percent of business owners believe employees value these office perks. But research revealed that very few employees, in *any* demographic, view benefits such as Ping-Pong tables (5 percent) or company outings (9 percent) as valuable to their workforce experience. More than half of respondents (53 percent) reported that having games in the office is a distraction and actually decreases productivity.

Implement new innovations that your employees will love. Your employees are no different from your customers: They want to see and experience new things that will help them do their jobs better. New tools, new strategies, new ways of solving problems—they want them! What they do *not* want is change that makes *your* job easier while making *their* job more difficult. Don't get the bright idea that asking your employees to fill out a new report or questionnaire is going to make them happy. It isn't.

Develop a center for employee happiness (CEH). No, I'm not talking about a refurbished break room with a Ping-Pong table and free bagels every Friday. Obviously, if you're a small business, the CEH is going to be located in your office. You will be the chief happiness officer (CHO) as you "manage by walking around." But in a big company, the salary you pay your designated full-time CHO will more than pay for itself in increased productivity and lower employee turnover and absenteeism. The secondary effect will be to produce fewer hatepoints and more lovepoints from your customers, who will respond to interacting with a happier, more engaged group of frontline employees.

Host employee listening sessions. These should be very casual, with small groups. Frankly, you should be listening *every day*, but on a regular basis you need to gather people together for an open discussion. What are we doing right? What are we doing not so well? Where are we behind schedule, and how can we fix that? The atmosphere must be nonjudgmental and honest, with low expectations—in other words, it's just as bad to overpromise as it is to be aloof and detached.

Create an enterprise social channel to collaborate around happiness. Digital platforms for meetings have come a long way in a short period of time. Enterprise social software is used by large organizations to improve their collaboration and social networking. Comprising corporate social networks (intranet) as well as other social networking software, the person-to-person connections it creates can boost the communication, productivity, and time management of a company. But while collaboration can often lead to fresh ideas and greater return on investment, it can also be tricky to pull off. You need to spell out who is in charge of what, and the tools you should be using to share the load.

Create and launch a happiness innovation challenge. An innovation challenge or hackathon is a way to involve the entire company or a group in generating innovative ideas or challenges that could positively impact customers or employees. Often the best innovations emerge through teams addressing unsolved problems. Creating a culture of happiness is no different. The structure of a happiness culture can often emerge from a "brainstorming" session, in which participants are encouraged to express what would make them happy at work, regardless of how crazy the idea might sound.

Develop and deploy a formal happiness as a strategy plan. This should be the culmination of everything that has gone before. It must include time-tested strategies as well as new innovations. It must include as many as possible of the ideas offered by employees that will help them be happy and productive, and create memorable touchpoints for their customers. It must be sustainable and supported by adequate investment

The plan should be deployed with measurements and be designed for continuous improvement.

Take Action!

✗ To create a culture of employee and customer happiness, it must be an enterprise imperative that comes from the board of directors, and CEO. Without the sponsorship and the philosophy of happiness, the path to enterprise happiness will fail.

✗ Your organization serves an experiential ecosystem, which includes customers, employees, vendors, partners, and your community. As you build out your happiness strategy, all participants within the experiential system must be served.

✗ Employee surveys are a plain and simple farce. Most employees believe that the survey will land in the hands of their supervisor, so they simply complete it in a way that protects their career pathway. Moreover, most surveys are constructed in a way that doesn't really get to the tough questions. Surveys can be used as a single date point, but rarely does the information from an employee satisfaction survey, or customer survey, provide insights that drive improvement.

✗ The best way to get accurate insights is with happiness hackathons. They create a safe place to share what employees love and what they hate about their job. The insights you gain when you use the right linguistics in a crowd setting is astonishing.

✗ Remember that your cultural ether is the life-support system for innovation, happiness, and ultimately enterprise success.

CHAPTER 9

Innovation and Employee Experience Design

Employee and customer happiness is an innovation activity that both improves your organization *today* and positions it for a brighter, more profitable *tomorrow*.

Let's start with a definition of innovation.

Innovation is the creation of *novel value that serves your organization and your customer.*

Let's dissect the anatomy of this definition:

- "Novel" simply means "new," and in a time of massive change you need to constantly find new ways to build new value for our customers and employees.
- "Value" is the tangible contribution you provide to the community or market you serve. It's something that helps people live better lives; gives shelter, nourishment, or health; or simply provides them with entertainment to brighten their day.
- "Your organization and your customer" is the human ecosystem composed of people who work with you and who receive your products and services. By delivering value to the customer, you

help them, but it's also good to create value for your organization in the sense that if the organization becomes bigger, stronger, or more efficient, it can better serve its customers and stakeholders.

Innovation—both enterprise and customer—needs a target, and the target of innovation for organizations are all of the humans that it serves. The better an organization is at coming up with ideas and deploying those ideas in a way that delivers value to the enterprise and the customer, the more likely is the success of the organization.

So now the only question is how you institutionalize innovation in a way that consistently targets employee happiness, enterprise growth, and customer satisfaction. This is an interesting concept because innovation means change, and so how do you then make "change" a permanent part of your organization's existence? Many leaders like to think that they need to make the organization a finely tuned machine that once perfected, will keep running indefinitely, and that if "something ain't broke, don't fix it." They say, "Making the business profitable the first time is a tough enough job, so once it is profitable, I'm not going to mess around with it!"

Instead of this static model, it's better to think of an organization like an airplane. Building it and getting it off the ground the first time takes a lot of work, but once it's airborne, the trick is to *keep* it airborne. You need to constantly adjust for wind and weather; while you may be able to put it on autopilot for brief periods, as a long-term solution that's not going to work. If you don't pay attention to changing conditions, you might find yourself crashing into the side of a mountain.

Your Employees Are Innovators

Your employees are natural innovators, and they want to have the opportunity to collaborate and cocreate. This is one of the most important ways in which you improve the quality and meaning of work life for employees. Leveraging internal social networks—sometimes referred to as enterprise social networks—you can set up innovation challenges and allow your employees to help you eliminate waste, improve customer experience, or anything else that you would like to accomplish. It can be done by collaborating in cocreating with your employees. The importance of this approach cannot be overstated, as employees want to know they're involved in the authorship of the systems, tools, and processes that they use every day in their work. We use the term "inclusive" a lot today, but the term is about allowing people to be included and that . . .

Innovation Creates Happiness, Which Creates More Innovation

Employee productivity is the backbone of any organization's success, and happier employees are more productive. And what gives human beings the most satisfaction and deep internal happiness? Creating something new that has value.

It's a virtuous cycle. Research has shown that human beings tend to become happy and derive satisfaction from activities that result in the creation of something new of value. Have you ever seen a child proudly show her new drawing to her parents or teacher? The child says, "Look! I drew a picture of a house and a sun in the sky!" Her happiness is palpable. We may grow up to be serious adults, but our delight in creating something new never leaves us.

The cycle begins when the happiness at having created something new then spurs us to do it again. Happy people tend to be more interested in exploring new solutions to difficult problems.

And we're not talking about geniuses being the sole sources of innovation. That's a myth. While there are some people who are truly exceptional, the vast majority of innovations come from ordinary rank-and-file employees. Reality shows us that under the right circumstances, anyone with normal human capacities is capable of producing creative work.

Another myth is that creativity somehow impedes productivity. Again, reality shows us that for complex work in organizations, there is no trade-off between creativity and the resulting productivity, efficiency, or work quality.

Harvard professor Teresa M. Amabile has done significant research on how the work environment can influence the motivation, creativity, and performance of individuals and teams. As she told *Harvard Business Review*, if a person is in a good mood on a given day, they're more likely to have creative ideas that day, as well as the next day, even if you take into account their mood that next day. When people are feeling good, a cognitive process leads to more flexible, fluent, and original thinking, and there's actually a carryover, an incubation effect, to the next day.

Not surprisingly, she found the behavior of the team leader is critical. She's identified four distinct leader behaviors that have a positive influence on the feelings—and therefore the performance—of subordinates. They are:

1. Supporting people emotionally.
2. Monitoring people's work in a particularly positive way, and giving them positive feedback on their work or giving them information that they need to do their work better.
3. Recognizing people for good performance, particularly in public settings.
4. Consulting with people on the team—that is, asking for their views, respecting their opinions, and acting on their needs and their wishes to the extent that it's possible.

Avoid These Seven Common Pitfalls

A happy culture positively impacts every aspect of your organization, your employees, and the customers you serve. A campaign to promote happy work and make it a vital part of company culture should be effective and produce results that you can measure.

With that being said, it's important to make certain you avoid the seven common pitfalls that can result from an organization's failure to achieve the robust cultural transformation it needs to sustain happiness day after day, year after year:

1. It's Just a Box to Check

Many employee happiness initiatives collapse because the CEO was never really committed to it in the first place! They look at "happiness" and "quality of work life" as a box they know they're supposed to check. This is one of the major reasons why ordinary efforts don't work. The CEO simply didn't commit the resources, time, and the philosophy of happiness to giving it a fighting chance.

2. It's Poisoned by WIIFM

Another common problem with attempts at cultural transformation is that organizations and leaders often view it from the perspective of "What's in it for me?" (also known as WIIFM). Happiness cultures are based on the philosophy of *serving others*, and that is the primary focus. Once you attempt to catalog the daily enterprise benefits without putting your stakeholders, employees, and customers first, your initiative simply won't work.

3. Ouch! They Stop Because It Hurts

Short-term pain is often necessary for long-term gain. Change of any type is painful at some level for a while. Transitioning from a traditional culture to one that puts happiness and humanity first is guaranteed to cause some disruption. Cultural transformation can be disorienting, disruptive, and even painful, and it often takes longer than one would expect. If you know this going into the initiative, you'll be far better prepared to stay the course in order to reap the amazing benefits.

4. It's Fractional

In the "vested ecosystem" (VE), everyone whom you and your organization will impact, and everyone within your ecosystem, must be served across their entire journey and across a range of personality personas.

Unfortunately, many organizations attempt to impact a small portion of their overall VE, and as a result they never really reach the goal of sustained happiness. Additionally, many organizations do not use the complete systems, processes, and tools that are necessary to drive a dynamic and happy enterprise, and as a result their initiative loses steam and ultimately fails. This really speaks to the importance of building out a formal happiness as a strategy plan that is hardwired to your entire enterprise strategy.

5. It's Monodirectional

Many organizations see all enterprise initiatives as something that they author and deploy onto their stakeholders. This "monodirectional" ("one-way") approach is flawed on many levels, and as a result it never works. One of the keys to employee happiness is the ability to consistently listen to and collaborate with your customers, employees, and stakeholders. Your employees want to be part of the discussion. They also want to be part of the decision making. Remember the best innovations that will help your organization thrive is in the heads of your employees, and if allowed, they want to share it with you.

6. It's Vague

One of the most important structures of a happiness initiative is the communication strategy embraced by the CEO. This strategy must provide ongoing, clear messaging on how employees can participate in the organization's journey to happiness and personal meeting. Unfortunately, many leaders see communication as incidental and a one-way broadcast. Building out a two-way initiative communication strategy as well as a CEO communication strategy is critical to taking the amorphous and squishy concept of happiness and making it an enterprise reality.

7. It's Contradictory

Here's a phenomenon that occurs too many times. A CEO talks about the importance of happiness and humanity, all the while creating policies and behaviors that are completely contradictory to those proclamations. (You can probably think of more than one well-known CEO who does this!) But you don't have to be perfect, and admitting your imperfection early on is a powerful way to get stakeholder adoption. Be honest with your employees, talk about what's not right, and then commit to collaborating and listening to them to create new humanistic approach toward building a fair, equitable, and beautiful work environment.

Take Action!

✗ Employee and customer happiness is an innovation
    activity that both improves your organization today and
    positions it for a brighter, more profitable tomorrow.

✗ Think of your organization like an airplane. Building it
    and getting it off the ground the first time takes a lot
    of work, but once it's airborne, the trick is to keep it
    airborne. You need to constantly adjust for wind and
    weather; while you may be able to put it on autopilot for
    brief periods, as a long-term solution that's not going to
    work. If you don't pay attention to changing conditions,
    you might find yourself crashing into the side of a
    mountain.

✗ Your employees are natural innovators, and they want to
    have the opportunity to collaborate and cocreate.

✗ Remember the virtuous cycle: Human beings tend to
    become happy and derive satisfaction from activities
    that result in the creation of something new of
    value. The cycle begins when the happiness at having
    created something new then spurs us to do it again.

CHAPTER 10

The Innovation Operating System

Your innovation operating system (IOS) is very similar to your computer operating system in that the goal of the system is to *run the machine*. An operating system (OS) is a set of programs on the hard drive that enables the computer hardware to communicate and operate with the computer software. It manages the hardware devices in your computer—things like the processor, memory, disk storage, keyboard, mouse, monitor, USB bus, and network adapter. Without a computer operating system, a computer and software program would be useless. A good computer operating system does its work invisibly, efficiently, and profitably. It's durable and requires only regular, anticipated upgrades to stay at peak performance.

Computer operating systems have names like Windows, Linux, Android, and Mac OS.

In addition to running the machine, computer operating systems are augmented by a range of specialized software packages designed to perform specific tasks. These software applications include things like word processing, spreadsheets, presentation software, and the like. Examples include Microsoft Office, Google Chrome, and Adobe Photoshop.

These are all commercial, mass-produced solutions. They are tried and true.

But your company is unique. An off-the-shelf computer software system isn't going to align with what you need. You need solutions that support your individual requirements.

That's why organizations are served by software vendors such as Intellectsoft, MojoTech, DataArt, and Frogslayer, who custom-craft software systems to meet their needs. The results are hybrid systems that have a basic architecture that's common to all, combined with features that are unique to the customer.

Your innovation operating system will be no different. It will have a basic structure that's similar to others in its class while at the same time having features that are unique to your organization.

This is *structured customization*. The key to a successful innovation program is that you need to follow a plan that ensures you avoid the common mistakes and pitfalls of enterprise innovation, while at the same time customizing it to fit the unique and special goals, needs, opportunities, and culture of your enterprise.

Create Your IOS Programming Code in Six Steps

All operating systems are constructed using code. The programming code is basically the rules that apply to the way in which the operating system controls the machine. All code is created using authoring platforms or languages.

Rather than diving too deeply into the analogy and going into all of the complexity of the language, examine the building blocks of your IOS code so that you can create an effective and efficient innovation infrastructure that will deliver real results to your organization.

You'll note that some of these steps dovetail with the work you've already done while formulating your innovation mission. That's fine. For example, the first step is to answer the question, "Why?" You may have figured this out and put it into your innovation mission. Good! Then you'll be able to breeze through it to the next step.

Let's get started.

1. The Why

If you've ever been around children, you probably know the story of the child who to every answer you give responds by asking, "Why?"

"I want you to come inside," you say.

"Why?"

"Because it's raining out."

"Why?"

"Because the clouds are full of water, and they can't hold it anymore."

"Why?"

"Because cool air can't hold as much water as warm air."

"Why?"

And so on. You get the idea. If you both keep playing the game, eventually you'll work your way back to the Big Bang Theory.

It's funny, but this shows why this question is both *important* and *challenging*.

It's important because it goes to the very heart of how the spark of innovation is vital to your organization. It's challenging because it takes mental effort to answer.

The why is critical because it drives stakeholder engagement, establishes measurements for success, and controls the ability to target and navigate through the entire process of innovation. Asking yourself *why* innovation is critical to your organization, and what you intend to do with it is truly the genesis of all innovation initiatives.

Perhaps the somewhat slippery and daunting nature of the question is why many executives rush to create an innovation initiative before they have asked themselves why they're doing it in the first place. But this is putting the cart before the horse.

If you're having a difficult time answering the question "Why innovate?" then consider this quote from Edith Widder, American oceanographer, marine biologist, and the co-founder, CEO, and senior scientist at the Ocean Research & Conservation Association: "Exploration is the engine that drives innovation. Innovation drives economic growth."

In her case, she's talking about literal exploration, as in getting in a submarine and going to the deepest part of the ocean to search for giant squids. Your exploration may be physical like hers, or it may take the form of searching for new ideas, processes, and business solutions.

Just as important, Edith Widder stated the second half of the equation: "Innovation drives economic growth." This is so fundamental you can print it on a big banner and hang it on your wall.

Exploring for new ideas drives innovation, which drives economic growth. This is absolutely true for any organization, including yours.

But to get back to how *your* organization answers the question, hosting an open discussion with stakeholders is a good way to start. Get people talking about it. Make innovation an ordinary topic of discussion on the same level as a sales report or new product rollout project. Remove the mystery and unfamiliarity. Make innovation part of the everyday fabric of your organization.

By the way, you don't need just one "why." In fact, you may have several. Here are some reasons why your organization—and your people—should embrace innovation. You can probably think of more!

1. Technology is constantly advancing, and our products need to keep pace.
2. Our customers have other choices from our competitors. We want to stay number one in the minds of our customers and keep them coming to us.

3. If we innovate, we can charge more for our products and make more money.

4. We want to hire the best talent. Smart, aggressive people don't like to sit around and do the same things year after year. They want to be challenged.

5. We're human beings, and we like to explore. Exploration leads to innovation, which leads to economic growth.

How many more reasons can you think of?

2. The Vision

From your answers to the question of "Why innovate?" comes your vision for innovation. It should be written out, like an organizational mission or vision statement that focuses on innovation.

And to answer the question, "Why should we have a vision for innovation?" here are three benefits of having a vision that's clear and concise:

1. **Engagement.** With a clear innovation vision, stakeholders will share a common goal and have a sense of being on a journey together. They will be less likely to waste time on nonproductive activities. They'll be more willing to accept the difficulties, challenges, and changes that the innovation journey can entail.

2. **Responsibility.** The innovation vision guides the innovation systems, which in turn manage the innovative pipeline and its various initiatives. With a vision and systems in place, staff can be empowered and given more leeway in their work. Because they know the goals and direction toward which they are working, they can be trusted to steer their own ship and determine the best way of getting there.

3. **Creativity.** If people know there are unsolved challenges lying ahead, they'll be more creative and willing to contribute more ideas. Because they've bought into the journey, they'll be more

motivated to find ways to go over and around the obstacles in their paths.

Here's what Satya Nadella wrote in 2014, a few months after he took over as CEO of Microsoft: "The day I took on my new role I said that our industry does not respect tradition—it only respects innovation. I also said that in order to accelerate our innovation, we must rediscover our soul—our unique core."

But lest you think Nadella was concerned only with big-picture, philosophical matters, later in his memo he got very much down to earth: "We help people get stuff done. Stuff like term papers, recipes, and budgets. Stuff like chatting with friends and family across the world. Stuff like painting, writing poetry, and expressing ideas. Stuff like running a Formula 1 racing team or keeping an entire city running. Stuff like building a game with a spark of your imagination and remixing it with the world. And stuff like helping build a vaccine for HIV, and giving a voice to the voiceless."

As with a strategic plan, the best innovation operating systems have the vision of innovation at the very top. This is important because a vision, being both intellectual and emotional, should be easy to communicate. You should be thoughtful in the creation of your vision statement, as this is something you will use to communicate to both internal- and external-facing customers. Just as you did when you asked the question "why?" don't go it alone! Make the process of forming your innovation vision open and transparent. Solicit stakeholder input. Generate a series of drafts, tear them up, and write some more.

The innovation vision must be powerful and transformative, and deserve the full support of every stakeholder. It must be fully embraced and championed by leadership. As former General Electric CEO Jack Welch said, "Good business leaders create a vision, articulate the vision, passionately own the vision, and relentlessly drive it to completion."

Remember that a vision statement isn't the same thing as a mission statement. While a mission statement describes what a company wants to do *now*, a vision statement outlines what a company wants to be *in the future*. It can also describe what kind of world it wants to help create for the future.

That being said, some companies blur the distinction. By whatever name you call it, the important thing is to *create it* and *communicate it*! Here are a few examples. I've added the emphasis.

"Bring inspiration and *innovation* to every athlete* in the world. (*If you have a body, you are an athlete.)"

—Nike

"To offer travelers a reliable, *innovative*, and fun airline to travel in Central America."

—NatureAir

"Our mission is to make Target your preferred shopping destination in all channels by delivering outstanding value, continuous *innovation*, and exceptional guest experiences by consistently fulfilling our Expect More. Pay Less. Brand Promise."

—Target

"We believe that we are on the face of the earth to make great products and that's not changing. We are constantly focusing on *innovating*. We believe in the simple not the complex. . . . We believe in deep collaboration and cross-pollination of our groups, which allow us to *innovate* in a way that others cannot . . ."

—Apple Computer

"Digital currency will bring about more *innovation*, efficiency, and equality of opportunity in the world by creating an open financial system."

—Coinbase

"At Philips, we strive to make the world healthier and more sustainable through *innovation*. Our goal is to improve the lives of

3 billion people a year by 2025. We improve the quality of people's lives through technology-enabled meaningful *innovations* . . ."

—Philips Research

"Offering all women and men worldwide the best of cosmetics *innovation* in terms of quality, efficacy, and safety."

—L'Oreal

"Using our portfolio of brands to differentiate our content, services, and consumer products, we seek to develop the most creative, *innovative*, and profitable entertainment experiences and related products in the world."

—Disney Corporation

"To be the most *innovative* enterprise in the world."

—3M

Because innovation is unpredictable, don't box yourself out of opportunities by being too specific with your vision statements. Organizations should, however, be big, brave, and bold in their vision statement and their innovation initiative alike.

Also, please note you can do your innovation vision anytime. Formulating a vision often takes time, so you may start the process and then continue it while you work through the other steps. And your organizational mission and vision statements should be reviewed once every few years to ensure they're still exactly what you want and are relevant. It's not uncommon for organizations to rewrite or revise their mission and vision statements as conditions change.

3. The Definition

Once you've determined *why* you're creating your innovation mandate and have a *vision* for your outcome or for the world as you hope to make it, you will then be well suited to begin the process of defining what it is and how it moves you toward your organizational goal. The definition should be understandable, relevant to your organization, and, most important, *measurable* and *attainable.*

Here's where a *gap analysis* is very useful.

A gap analysis involves the comparison of actual performance with potential or desired performance. It involves determining, documenting, and improving the difference between business requirements and current capabilities. It's a formal study of what a business is doing currently and where it wants to go in the future.

Please note that the capitalized "GAP analysis" has also been used as a means of classifying how well a product or solution meets a targeted need or set of requirements. In this case, the acronym GAP can be used as a ranking of "good," "average," or "poor."

In everyday business, a gap analysis can be used to define what it will take to meet a goal. Here's a simple example:

1. Identify the status quo: We're selling 5,000 units per month at a net profit of $1 million.
2. Identify the target: We want to net $1.5 million within two years.
3. How can we reach the target?
   A. Cut our expenses so we can sell 5,000 units but net $1.5 million.
   B. Sell 7,500 units of the same product at the same price.
   C. Raise the price per unit by selling an innovative model with more features.
   D. A combination of A, B, and C.
4. Challenges.
   A. We believe our existing geographical market is saturated, and we're not ready to expand into a new market.
   B. Our factory is at 100 percent capacity and we can't expand before two years.

5. Strategy: We will seek to bridge the gap by *innovating* and making our product a *better value* so that customers will pay more for it.

You can define what you want to get from innovation and set appropriate and attainable goals. Here are just a few examples:

- Mandate that a certain percentage of annual revenues must come from new products. This is often called the "innovation sales rate" (ISR).
- Track how many ideas per month you're getting from all of your employees.
- Measure the success of individual innovation projects (from concept to customer) and overall platform or new business development programs.
- Calculate the risk-adjusted net present value of the innovation pipeline and the return on investment in that pipeline. We'll talk about the innovation pipeline in the pages ahead.

Remember that inherent in innovation is exploring the unknown, and that brings with it an expected rate of failure. Accordingly, it's important to measure innovation holistically. Each individual effort cannot and should not be measured at the innovation state. Beware of measuring only what's *easy* to measure instead of what's *important*, and avoid measuring too many things.

4. The Readiness Assessment

Innovation is a people-powered process. Innovation is literally the process of gaining insights and ideas and putting them into action, and as of today, robots and computers aren't good at this. People must do it, and they need to be ready and willing.

Complete your assessment to determine how you will reach out to stakeholders, partners, vendors, customers, and in some cases even competitors, to get insights that you can transmute into organizational and customer value.

A gap analysis isn't just for the operational stuff. It can also be useful when looking at the readiness of your leaders and employees.

For example, the Korn Ferry Institute conducted a side-by-side comparison of a group of logistics executives from average companies and those from *Forbes* magazine's "The World's 100 Most Innovative Companies" (MIC). The study was based on the premise that if innovation were crucial to successful strategies across a company's supply chain, then logistics sector leaders needed to possess the personal qualities that would enable the spark of innovation to flourish.

The study found that when compared to their MIC peers, the group of average logistics executives "typically displayed lower levels of learning agility and cultural dexterity—two traits highly predictive of success and engagement, especially in senior leadership roles."

Learning agility is defined as the willingness and ability to learn from experience and then apply this knowledge to succeed under new or first-time conditions.

Cultural dexterity is defined as a professional's ability to work effectively with individuals from various cultural backgrounds.

Both are important for sustained innovation.

The Institute reported the average logistics leaders excelled in only one among eleven innovation indicators—independence. Meanwhile, executives in the *Forbes* MIC received top scores of ten in seven areas: cultural dexterity, learning agility, emotional intelligence, self, power, challenge, and independence, and scored nine in thought, which has a specific innovation component.

Overall, the *Forbes* MIC executives outperformed the average logistics leaders by nearly 50 percent.

This is not surprising!

The embracing of innovation must be well-thought-out and sustained.

Here's some tough love: Stocking your employee lounge with Ping-Pong tables, installing whiteboards, hosting quarterly hackathons, and proclaiming casual Fridays isn't the way to ensure innovation readiness!

Leaders often seize upon these quick-fix tactics and convince themselves they are ready for innovation when, in fact, the critical ingredients of innovation are missing.

Instead, survey your managers and employees and ask them if they can define the *problem* they're solving for the customer. You don't want them to describe the current solution. You don't want them to say, "We make a great product." That means they're only thinking about today. If they're focused on the *problem*, they'll be thinking about how to create better solutions for the future.

When your people are ready for the spark of innovation, they'll capture it and harness its tremendous energy.

5. The Team Architecture

Your innovation mandate needs to be both well-defined and flexible.

Strong but able to bend.

Identifiable but shape shifting.

And while it needs to be democratic and woven into the fabric of your organization at every level, like any other operational function it requires guidance and oversight.

While everyone in your organization should *participate* in innovation, you don't need an organization full of innovation directors. It's no different from, say, quality control. While everyone in an organization should support and pursue the highest standards of quality control, you don't want them all to be quality-control officers. You need people who can oversee and evaluate your innovation program, just like any other operational area of the organization.

If your organization is large, you may have multiple innovation teams. Most teams will appoint a team leader, who is responsible for establishing process, including how to communicate during brainstorms and meetings. That person can also help guide to the top of the organization the ideas and initiatives introduced in those meetings.

At the very least, you need an *innovation champion*. In a small company it could be the CEO, or in a larger organization there might be several. The innovation champion has the power to allocate resources within a set budget and ensure that the innovation pipeline is full.

Continuous Improvement, Big Breakthroughs . . . Or Both?

Innovation generally manifests itself in two ways.

1. **Continuous improvement.** This is the Japanese *kaizen*. Here, you're looking for a steady stream of small ideas from front-line workers in every department, which when implemented will deliver an incremental improvement to a process or product. To create and maintain this kind of approach, the only significant investment you need to make is in your *people*. They need to know they're *expected* to offer ideas, they need to know *how* to do it, and there needs to be a person or people who *receive* the ideas and process them.

   Hackathons are fine, but only as part of a long-term strategic embrace of innovation. If a hackathon is just a one-shot event, the end result will be worse than doing nothing, because the employees will feel used.

2. **The big breakthrough.** This is what we see in structured innovation programs that have a specific project goal, such as to find a new drug to combat a particular disease. Big breakthroughs require line-item funding, staffing, and often a capital investment in space and equipment. The company looks for a return on investment, not necessarily from each specific project but from the innovation effort as a whole.

   Many companies use a hybrid approach and look for both continuous improvement as well as big breakthroughs.

Your People Are the Key

No matter how you organize your innovation effort, you must have people who are responsible for the entire innovation process.

Here are the key attributes of people you need on your innovation team:

- You need an innovation champion on the team who has access to the "go" button. This is absolutely key. If new ideas can't be propelled across the finish line, team members will quickly figure out they're nothing but empty window dressing.
- In architecting your innovation team, it's incredibly important that you leverage a thoughtful group of people who have the ability to provide real value in the assessment and development of innovations. You need subject experts in the relevant areas— marketing, design, production, human resources, finance.
- This is not about putting together a group of sycophants; in fact, you want a team that will likely get a bit scrappy, given their wide range of views on all things innovation. Bring in a mix of veterans with deep experience of the core business but who may resist seeing new possibilities, along with younger "creatives" who are wired to generate crazy ideas that can't be implemented. Make them work together to recognize the spark and capture its full energy.
- The team members should be results-oriented, fast-moving, smart, and perhaps most important willing to take some big risks. You don't want people who are comfortable simply punching the clock day after day.
- When working on a project basis, an effective group is comprised of individuals from different departments within the company. If the team's goal is to develop a new product, for example, there should be representatives from research and development, design, manufacturing, marketing, sales, and finance on board.

6. Coding Your Innovation Operating System

Now that you have a reason *why* you need innovation, a *goal* that you want to achieve, and a *team* that will help you get there, it's time to build your plan.

Thousands of multibillion-dollar corporations, not-for-profit corporations, and universities have created innovation mandates. They all have the same structure when it comes to innovation, which is about helping an organization meet its organizational mission by leveraging novel ideas that are valuable. Chaotic changers successfully use the following structure in the creation of the best innovation operating systems.

Establish a Clear Team and/or Organizational Process

Whether for a project team or the entire organization, a clear process ensures every employee can both participate in innovation and be held accountable to the same principles. Establish a leadership structure, define the roles of each team member, outline norms for how the team should collaborate, and set individual and group goals.

The more clarity and empowerment you can give the team and overall organization, the better.

Allocate Sufficient Resources

Like every functional area of your organization—human resources, operations, finance, compliance, logistics, marketing, or any other—innovation needs resources that will always include *time* and may also include *space*, *equipment*, and *information*.

Don't make innovation an also-ran. Because most team members aren't working solely on a single innovation project, employees are often pressured by other work demands. This is where the innovation champion is particularly useful—to give employees official permission to take chances or explore new ideas.

Encourage and Embrace Failure

Innovation and failure go hand in hand. That's to be expected.

Failure should mean that the team tried a new idea and learned an invaluable business lesson. For employees to believe that failure is okay, the organization needs to create a climate of "psychological safety"—a term coined by Harvard Business School professor Amy Edmondson. It's a work environment in which people feel comfortable admitting to well-intentioned mistakes without fear of being punished. What no one wants is a climate where mistakes are made and then covered up because team members don't want to be made to feel vulnerable.

Failure should be written into the budget. Unless a company is defined by its breakthrough products, the company's bottom line should never wholly depend upon the success of one risky project.

4. Ensure Employee Engagement

When innovation is a team or project effort, leaders need to ensure that everyone understands and supports the organization's commitment to innovation and that the team receives the cooperation it needs to succeed.

While in a large organization it's neither necessary nor desirable to plug every employee into the decision-making process, the more visibility innovators can bring to their work, the more willing other stakeholders will be to help and adopt new services, processes, or tools.

Making everyone comfortable with innovation is best done through personal conversation. Leaders need to lay out the vision, take questions, and get sense of the organization's appetite for both incremental innovation and larger risks.

Before trying to make the big sell, innovation team members can gain a better understanding of the problems facing the organization and uncover any main points of objection they might face by meeting with other colleagues as well as the innovation champion.

Innovation is a series of concrete, definable actions. You don't have to be born with the ability to be innovative. Because it's a set of behaviors that can be learned, anyone is capable of doing it—that is, as long as the organization makes it a priority.

Take Action!

To create your innovation operating system, follow the six steps:

1. **Why.** This is your compass that keeps you pointed toward your goal. Do not start until you know exactly why innovation will benefit your organization.
2. **Vision.** "See" the innovations you want to create. Not the innovations themselves, of course, but the problems you want to solve and how your organization could make a difference.
3. **Definition.** What kinds of innovation are you looking for? Incremental or breakthrough, or a combination? Process or product? Remember, innovations don't have to be directly experienced by the customer. For example, significant innovation is happening in logistics and human resources.
4. **Readiness.** Innovation is created and championed by people. Your leaders and your employees must be ready and willing to embrace innovation in all its forms.
5. **Architecture.** Do you need defined teams to work on projects, or are you looking for every employee to offer ideas? Be sure to identify—and empower—as many innovation champions as you need.
6. **Coding.** Establish the structure of your innovation program, with clear procedures and processes. These need to be both structured and flexible. Ensure buy-in from every relevant stakeholder.

CHAPTER 11

Three Simple Steps to Driving Innovation

Here's a powerful three-step system for driving innovation excellence.

Step 1: The Innovation Readiness Assessment

If you went to your doctor and she prescribed a heart medication without doing any tests or examining you, you would justifiably be suspicious. How about your mechanic? You bring your car in for service, and without even looking under the hood, he claims you need a new transmission. Unless the doctor or mechanic has some sort of X-ray vision, you would turn around and walk out.

The same is true of organizations. You cannot set up an innovation program without first doing a thorough assessment and diagnosis of the problems you are trying to solve and what is currently being done about them. To determine if your organization is ready for an innovation program and what it should look like, you need to start with a good assessment.

Surprisingly, most organizations begin work on innovation without doing a complete and thorough self-examination. Remember, innovation is a delicate ecosystem consisting of hundreds of moving parts, including customer types, expectations, sensory input, price/value sensitivity, collaborative environments, and so on. To succeed, you must have the systems, methods, tools, and processes necessary to make innovation real in your enterprise.

Prepare for Your Readiness Assessment

Sometimes you can accomplish a task on your own, and other times you've got to have expert help. You'll probably need an outside expert to conduct an innovation readiness assessment (sometimes called an innovation gap analysis).

For most organizations, it is critical that the assessment be done by a company that specializes in innovation best practices. These companies have trained experts with an outside perspective.

Conducting your own innovation readiness assessment is kind of like doing your own tonsillectomy. It's going to hurt and you're not likely going to have a good outcome. But should you decide to do your own assessment, first get training on innovation best practices to get a good perspective on what's really required. Meaningful customer training is essential because none of us is born an expert in customer experience; we need to develop the skills. Most organizations hire really great people to do customer service but then shove them out without any training. All training should be customized to the unique needs of your organization and should be specific to the job title; for example, sales staff should get one type of training, while executives get another.

On the other hand, so-called innovation gurus are popping up everywhere, so make sure you choose wisely. The last thing most organizations need is to bring in a new level of bureaucracy, because bureaucracy is the enemy of innovation. Hire a company that has experience mopping up failed initiatives, and avoid any company that has a just-add-water solution. No one-size-fits-all template will work.

A watchmaker once talked about watches (and clocks) as delicate eco-systems. With any slight variation in mechanical tolerances, two things could happen, both bad:

1. The watch would come to a complete stop. Actually, he proclaimed, "Stopping is good," explaining that when a watch stops, it's telling you that it's broken, and you are prompted to get it fixed.

2. The watch would continue to run, but unbeknown to the owner, it would no longer be accurate. The owner would assume all was well, and therefore the watch wouldn't get repaired until the owner missed an important appointment!

And that's where the danger lies. Innovation is like a watch or clock. It has hundreds of moving parts. If these moving parts are not all working in synchronicity, your innovation ecosystem collapses.

Many organizations have ticking watches and therefore assume they are being innovative. For example, a company's new website might be attracting online customers, so the company incorrectly believes its entire innovation system is optimized and the information gleaned from the market is correct. The company assumes the methods of developing innovation are ticking along nicely, only to find itself blindsided by an existing or emerging competitor.

Most people would never dream of trying to tinker with their watch. It's too complex and requires an outside expert who understands the hundreds of moving parts and how to optimize them. Innovation programs also have moving parts that come together in a formal innovation governing structure. Innovation programs can be complex. Most organizations execute an innovation without going to an expert. Such organizations often omit the diagnostic and developmental stages and are not optimized through innovation best practices. Yet this is exactly what's needed for best results. For this reason, proper innovation is typically designed and implemented using outside experts.

Step 2: The Innovation Road Map

Based on your innovation readiness assessment, you've determined that you have the correct resources, systems, and talent in place to get started. You also have determined where your innovation gaps are.

The next step is creating your innovation road map. Surprisingly, the overwhelming majority of organizations looking to drive enterprise collaboration and innovation never actually create an innovation plan. "Failing to plan is planning to fail," as the saying goes, and this is certainly true when it comes to innovation.

If you were planning to hire a vice president of finance, you would first carefully assess her qualifications, training, and experience relative to financial control and leadership. Then you would have detailed discussions about where you wanted the company to head and how you envision getting there.

Yet most stakeholders charged with the responsibility to execute on innovation have little or no training specific to innovation best practices. Building a road map is a key element of those best practices. This is why it's so valuable to hire or have trained stakeholders who understand the working parts of innovation best practices.

You would not allow someone who just "felt" he was a natural born surgeon perform even minor surgery on you! You would probably run—fast and far—in the opposite direction. Surgeons must complete years of training to make certain they know what they are doing. Innovation is no different. It requires specific learned skills. Organizations that do not provide innovation training and coaching are doomed to fail.

Design an Innovation Road Map

Once you have completed a comprehensive innovation readiness assessment, do yourself a favor and take the time to build out a road map that is customized to the uniqueness of your organization. Make sure not to skimp on how you build it. Good innovation initiatives provide significant and predictable returns on investment.

Studies show that the main reason innovation initiatives fail is that they were not surgically connected to the enterprise strategy. In designing your innovation road map, make certain you're using innovation as a way to serve your stated enterprise goals.

Step 3: Innovation Execution and Measurements

In medicine, the correct medical pathway is for the doctor to perform a thorough assessment and diagnosis, develop a focused and quality treatment plan, and ultimately deploy and measure the treatment and results.

This linear, step-by-step approach makes great sense. It is also the perfect process for winning in innovation. Once you've completed your innovation readiness assessment and developed your comprehensive innovation road map, the final phase is execution and measurement. For this, you must determine how you're going to measure success, fill all of your resource gaps, and launch with sustainable determination.

Different sources quote different ways of measuring innovation success. Some have three essential ways, while others have ten. For the sake of simplicity, try measuring innovation success by determining how much it helped you improve a specific strategic initiative. How much better are you as a result of using that innovation best practice? For example, if your goal was to reduce cost, by how much was it reduced? For our purpose, we're talking about how much an innovation has helped you improve the quality of the experience you've delivered to customers across their journey.

Innovation is a lot of fun and is naturally connected to the state of being human. We are creators. That's what we do in the process of deploying innovation, and it's incredibly exciting and rewarding.

The Six Cs of Innovation Success

A study of successful innovation initiatives will reveal there are six elements, or the Six Cs, that must be in play for any innovation program to succeed. Here are each of the six elements needed to bring a successful innovation program to your organization.

THE 6 Cs OF INNOVATION

Complete. Ninety percent of all innovation programs fail because they are incomplete. The innovation ecosystem is a delicate one. The various species of success must all be alive and well or your ecosystem will collapse.

In other words, you need innovation governance, training, systems, processes, tools, and technologies in place and ready to go. Leave something out, and innovation becomes nothing more than a logo and a corporate buzzword. Your innovation readiness assessment, done properly, will make sure you have everything you need to be successful.

Customized. Every industry is different, as is every organization within that industry. Run from the purveyors of the one-size-fits-all solution. Innovation has to be handcrafted to align with your organization's appetite for risk and opportunity and your specific and unique customer types.

It also has to be culturally aligned and realistic in terms of your objectives and how you measure success. By doing a complete innovation readiness assessment, you gain the insights necessary to know how to customize your innovation initiative to fit the unique and special needs of your enterprise and customers.

Culture. This is the life support system of innovation. If your organization is risk-centric and fears collaboration internally and externally, your innovation program is doomed. If you have a fear-based, noncollaborative culture, then chances are you have many other problems related to enterprise success.

Today, organizations need to attract millennial talent. They need to cocreate and collaborate, and ultimately they need to innovate. To do that, you have to be willing to "encourage courage." Get your culture right because it's the new enterprise mandate.

Collaborate. Very few innovators have created anything completely on their own. One way or another, people worked together to create that new product or service.

Most people fear collaboration because they're afraid someone else will take credit. In some cases, they're afraid somebody will outright steal their idea. But in fact, the best innovators are the ultimate collaborators.

Develop collaborative environments within your organization, create places and times for people to exchange ideas and innovate, and ultimately make collaboration part of your organization's culture.

Connect. Being connected is an often-used phrase, but in most organizations, people rarely have important relationships with others. However, innovation and collaboration require being connected. One of the best ways to stay connected is through technology, such as enterprise social networks and similar platforms.

You can also conduct brainstorming sessions, create innovation labs, and have other ongoing innovation activities. Deploy these tools to get creative people together sharing ideas and experiences and to help find more opportunities to invent exquisite customer experiences.

Customer-Centric. Essentially, there is a philosophical difference between organizations that master customer service and experiences and those that fail to do so. Innovation success derives from customer-centric organizations; innovation failure stems from those that are company-centric.

It's paradoxical, but the more you focus on delivering internal enterprise benefits, the less you focus on your real enterprise benefit, which is a loyal customer. The more you focus outward, on your customer, the more inward success you eventually have.

Take Action!

✗ Remember, whether you're selling a product or a service, you are in the customer experience business. Customer experience is not just a function of training your staff; it's a design function. If you want to design something that's going to succeed, it needs to deliver value to customers across each and every touchpoint.

✗ Innovation is as much a philosophy as it is a business discipline. The philosophy begins with a customer-centered view of the universe. A fractional approach will result in failure, and even worse, you will lose credibility. Don't deploy on innovation until you really mean to carry it through, and when you do, make sure you do it right. When you launch innovation initiatives, build in dashboards and measurements so you can see what's working and what's not. Build a team culture that encourages courage. Put structures in place that help you gain better insights from your customer-facing stakeholders. Beware of "just-add-water" innovation programs, because there is no such thing as a successful cookie-cutter approach toward winning at innovation.

✗ Make your innovation initiative a beautifully handcrafted, custom program that fits the unique and special goals and culture of your business. Don't forget, this is really a lot of fun, so make it buoyant, engaging, and worth your team's effort.

CHAPTER 12

Technology and Customer Experience

Technology as it relates to customer experience can be a double-edged sword. On one side, many organizations lean on old-fashioned customer relationship management tools and digital surveys. Even though they usually provide erroneous and fractional insights about what customers truly love and truly hate, installing *some* form of technology gives many executives a warm, comfortable feeling. It's as if doing so removes the responsibility of really getting to know and understand their customers.

On the other side of the coin, there's no doubt that good technology can provide better insights and ways of aggregating and responding to those insights. But if this is done without the proper context, it can result in failure.

Every organization should develop an integrated power plan that includes *both digital and nondigital channels*. It should also include technology stacks, which are ranges of technologies that help you do a better job of identifying what customers need and want and how to deliver better solutions to them. In this way you can use the powerful insight-gaining potential of technology for the right purpose, which is to deeply understand what our customers love and what they hate.

Without a complete and integrated plan, you risk leaning on old-fashioned, fractional approaches toward customer experience. We will assume that customer relationship programs—typically marketing programs—are going to save the day. Of course, as we've seen too often, they can't.

Technology as Power

The old-fashioned methods—often CRM programs—of leveraging technology were really about trying to find new ways to push out sales pitches to customers. The problem with that is customers don't want to be managed. They want to be honored and respected.

So as we begin the process of leveraging technology, including AI, we do it from the perspective that the technology we use will enable a more efficient and effective job of delivering surprising value to the customer across both digital and nondigital channels. If we start from that perspective, the insights gathered will be far more relevant and will move us toward innovation and customer experience superstardom.

Five Key Drivers of Technology for the Long Term

Here are the five key drivers that will determine the success of your enterprise as it relates to technology. Regardless of changes in the foreseeable future, these drivers will remain relevant, and they will be applicable to the way you build your customer experience power plan. They will affect the use and benefit of technologies over the next several decades.

Driver #1: AI Ubiquity

One of the most conspicuous trends that will change the way in which customers engage us, the way we engage them, the way we gain insights, and the way we measure and monitor the results of the products and services we deliver to the market is the concept of AI ubiquity. This is the idea that AI is everywhere, and it will only become more so.

The way in which we leverage the advantage of being continuously connected to our customers—and seemingly able to outguess them—will have a massive impact on the way we deliver and build delicious customer experiences. AI ubiquity says our customers can find us, research us, vet us, and try to glean value from us, and we can do it back to them. They're engaging our products, services, and brand through connected devices, and vice-versa.

Mobile devices creating micro-digital moments (or micro-moments) are incredibly important in terms of how a customer engages and connects to a brand. Tomorrow, this digital ubiquity will only expand as our connected devices become ubiquitous wearable technologies.

There is digital consistency today in that we're all connected. In the next wave, we will of course still always be connected, but we will be connected to a great many more things and in more ways than we are now. We will be in touch with our customers, they will be in touch with us, and we will have the advantage being able to glean data and insights as never before.

So how is your organization going to leverage the fact that your customer today is digitally ubiquitous—and will be even more so in the future—and constantly connected? What are you going to do to leverage the power of AI and create amazing experiences? We must create seamless, integrated, and elegant connections blending both digital and nondigital human experiences.

Too many organizations try to create dual customer experience silos: (1) the real world, nondigital, side, where team members build real-life experiences for their customers, and (2) the digital side, where the team builds out digital experiences. Regrettably, organizations often do a poor job of seamlessly integrating the two.

The trick is to always build out these experiences concurrently with your entire team to ensure that the products, services, delivery channels, and branding square up to create one integrated solution. In this way, your digital ubiquity and your digital brand promise will be in sync with the nondigital physical, real experiences you also provide.

Driver #2: Granularity

In the past, our tools for obtaining consumer data were poor. Often, the information was so vague that we couldn't use it to come up with customer-driven innovations. Fortunately, today we can be extremely granular. We don't just look at pet owners, for example; we can look at female owners of French bulldogs who live in Brooklyn.

The more granular we become, the better we will be at delivering relevant messaging, and the better job we will do of delivering products, technologies, and services that the customers we serve consider excellent. Technologies will continue to afford us the ability to have far greater granularity in the way in which we identify our customer types, and because of this, we will be able to build packets of solutions that are very, very relevant to them.

Remember, historically we were designers of macro-customer experiences. Today, we must be micro-designers, designing micro-experiences to micro-segments of a subsegment. It may seem ridiculous to get that granular, but through technology we have the ability to do it, and because we have the ability to do it, we should do it.

The headwater of excellent customer experience is really relevance. If we want to create relevant experiences, we must become more granular in understanding the unique segments and subsegments of all our customer types so we can build out messaging and solutions that are special to them.

Driver #3: Meaningful Data

The third trend is meaningful data. Cognitive computing, a subset of AI that is typically used to describe AI systems designed to simulate human thought, is one of the most exciting areas in big data and data analytics. Not only must we get lots and lots of data (because, well, we can), but we also need to understand what that data means. Every data signal has meaning. We need to look at every piece of data and leverage the power of cognitive computing to understand ways to deliver better experiences. Many times this will happen in ways we never thought of.

The future of technology, as it relates to customer experience and to meaningful data, will have much to do with acquiring large amounts of data sets. Cognitive computing will take that data and do a far better job of aggregating and leveraging it to understand its full meaning. Once we begin to see these small signals, we can identify unique and special ways to deliver perfect human experiences.

Social listening is another exciting area in big data and data analytics. Analyzing the voice of the customer through social media data provides organizations with the ability to understand consumers as they never have before. Consumers share hundreds of billions of posts about their experiences, likes, and dislikes every year.

Customer experience and marketing executives can not only use social data to measure brand health and the impact of marketing campaigns on purchase intent, but they can also use this data to identify unmet needs in the market to inform future product innovation.

Using cutting-edge big data analytics techniques, you can perform virtual ethnography at scale. Organizations that tap into unsolicited social data have a distinct advantage over those that rely only on traditional, solicited data (such as surveys) that frequently merely confirm what they already know, thereby perpetuating the status quo.

Driver #4: Actionable Insights

Actionable insights derive from having meaningful data; this means having information we can actually do something with. Seems pretty obvious, but companies have spent millions of dollars to obtain insights that were all one-dimensional and meant virtually nothing. In the future, the beauty of technology—the elegance and poetry of technology as it relates to customer experience—will be our ability to understand the meaning of what our customers are doing, what they're saying, and how they're behaving.

When we are able to understand the meaning of our customers' behaviors, we have the nucleus of what we need to create disruptive innovations that will blindside our competition while delighting our valued customers.

Driver #5: Measurability

The last trend in the future of technology is measurability. Heed the familiar axiom: "What gets measured gets done."

Sadly, too many organizations are involved in initiatives that either can't be measured or use incorrect measurement tools. Therefore, most organizations have little ability to determine the success of a range of initiatives. Worse, some organizations measure only the success of an initiative based on its profitability, without regard to its impact on the customer and the market the organization serves.

Measurements should be in the form of executive dashboards, which allow you to take lots of complex data and display it in a simple graphic way. It allows you to clearly see the effect of current customer experience initiatives and pinpoint ways to improve your customers' experiences based on the data. Setting up a wide range of powerful measurement tools and reporting them to key executives via executive dashboards is by far one of the most powerful technologies that will serve the future of customer experience, particularly as the data and methods by which it is delivered will continue to improve in the future.

Most organizations look at the way they impact customers annually or—surprisingly—they never do it. However, by using executive dashboards that are in the hands of people who have the authority, inclination, and motivation to make changes quickly, you can make sure your company doesn't go off track. Your organization can be proactive instead of reactive. In our hypercompetitive environment, by the time we realize we're headed down the wrong road, it's often too late. We've been blindsided by the competition because we didn't have the pulse of our market, our customers, and their entire journey at our fingertips.

Executive dashboards leveraging cognitive computing and better insights across the customer experience are the future of customer experience.

A word of warning: Don't use technology solely as a shortcut to drive profit. Technology can increase efficiency, reduce costs, and build sales. But if this is your principal focus, chances are you'll be delivering mediocre customer experiences. You can't develop an awesome customer experience or an organizational culture that is customer-centric just by using technology! Don't fall into the lazy person's trap of plugging in a technology stack and waiting for your happy customers to give you a 5-star online rating. It won't work.

Take Action!

✗ These five trends will morph in many different ways. Some of them are in development stages and some are already available.

✗ AI ubiquity is here and will continue to be more so as new technologies are invented and AI continues to evolve.

✗ Granularity is the only way we can create real and meaningful experiences across a wide range of customer experiences.

✗ Meaningful data, understandable information about what our customers really care about, is essential. The key to managing lots of data is the ability to aggregate it and identify what matters.

✗ The ability to come up with meaningful data through cognitive computing and social listening will have a major impact on the way we identify meaning. It will allow us to have actionable insights to transform what we learn into innovations that deliver exceptional customer experiences instantaneously over long developmental life cycles.

✗ And finally, measurability via graphic user interfaces or executive dashboard interfaces will allow us to get complicated information reported to a range of executives quickly so that they have the authority, motivation, and inclination to move fast and improve the quality of our customer experiences.

✗ These five technology drivers should be part of everything you do as you build your power plan. There is no doubt technology will have an incredible impact on the future of the customer experience.

CHAPTER 13

Top Ten Innovation Killers

Organizations are made up of people. These people have habits, and they have beliefs. Psychologists have taught us there is one way to change bad habits—through what is referred to as "meaningful consequence."

In my practice, I've found that people don't come to us until they've tried every "grapefruit diet" and every magic pill. It isn't until innovation has stopped, competition is killing them, and the organization has started to fail that they develop the lucidity to recognize that things have to change.

Innovation is an active process that may require deep organizational change. There may be pushback from people who don't want to change or systems that are not yet robust enough to handle it. As you embrace innovation, be on guard for these top ten innovation killers—and take steps to vanquish them!

1. The Wrong Focus

Innovation success can only be driven by a continued and systemic focus on delivering *true customer value*.

Organizational focus is key. Most organizations that fail in the area of innovation fail because they look at it from a departmental, or "siloed," approach. They look at it fractionally. They look at the speed of the internet or the "cool" of the internet or some other single feature of it that will solve all their problems. It doesn't. Just creating a website for your customers doesn't mean you're creating great customer service. Again, it's like the "grapefruit diet"—it's applying one aspect of what should be a holistic innovation process and assuming that is going to solve the problem.

In order to win at innovation, you have to create real meaningful net customer value, that is, something your customer genuinely values and is willing to pay the going price for. It means the thing you deliver to them can't just be a feature; it must provide a benefit. So many people in the innovation space create "pseudo-value" because they are creating benefit statements that are really feature statements, and those features deliver no value whatsoever.

2. Lack of Sponsorship

Without sponsorship, you'll never create the insanely cool products that will deliver meaningful net customer value.

Successful companies know that they need to secure innovation champions, who will take ownership and the responsibility for successful commercialization of products that need to succeed in the marketplace.

Sponsorship can take many forms, and it can occur at many levels in an organization. The most prominent, and often the most successful, sponsorship efforts are not always the most visible. But when they are visible, with news coverage in the public or trade media, it's usually a signal of something larger. Prominent examples include Toyota's declared "Value Innovation Strategy," a program launched to bring greater focus to product innovation in the wake of the Prius model's success. The company, to that point, had been more renowned as a manufacturing process innovator. The Value Innovation Strategy was at once a mantra and evidence of sponsorship for product innovation. It worked not only inside the organization as a wake-up call for more product innovation and to "see the big picture," it also opened the door to suppliers to participate in the innovation process beyond basic cost-cutting tactics. Although some may place blame for Toyota's recent recall troubles on such expansion; in my mind it represents a clear break from a process-focused past.

BMW gives us another auto industry example. Each time the company begins developing a new car, the project's team members—about 200 to 300 staffers in engineering, design, production, marketing, purchasing, finance, and so on—are relocated from their scattered locations to the automaker's Research and Innovation Center, called FIZ, for up to three years. The close proximity speeds communication, enables better communications, prevents unforeseen conflicts, and more—but it also sends a strong signal of innovation sponsorship straight from the offices of top management.

3. Process Driven to the Hilt

Another major problem is "process driven innovation." Successful innovation is not about the best process. The current hyper-focus on the innovation process is, in fact, killing innovation. Winning companies use processes to build speed and improve customer value.

Innovation "losers," on the other hand, focus on internal needs and risk management that results in subincremental innovation—and oftentimes, a technology that is not customer or market relevant.

4. A Risk-Centered Process

In a world of chaotic change, a constant daily focus on managing risk at every level will increase the risk that a company will *lose* its competitiveness and market share. Obviously, the introduction of great innovations, and the launch of great products, is a good way to validate the innovation process. But the surefire way to avoid risk "drag" is to build and expand the conduits of connectivity to your customer to ensure, or at least increase the odds, that the products you deliver are meaningful and relevant.

"Hyper" risk management kills innovation, reduces speed to market, and effectively eliminates breakthrough market innovations.

It also leads to an obvious symptom: incrementalism. This is one of the biggest problems—or symptoms, really—in the area of innovation today. Incremental innovation is not so much a problem in and of itself, rather, it is symptomatic of a greater problem—or greater problem(s)—most of which center on the common organizational fault of risk aversion.

Really great innovations often look crazy. But really bad innovations can also look crazy! Great ideas and bad ideas both look strange, and most people don't do anything with them. So what do we do instead? We paint it blue, and call it new.

Incrementalism may justify the research and development function, but it doesn't build meaningful value for the valued customer or for the organization. Eliminating incrementalism has a lot to do with dealing with its root causality—and that's risk aversion. If we're afraid to do something big to deliver impactful customer value, we make incremental improvements to the existing technology, and at the end of the day, that is a failure.

5. Customer Be Damned

How would you like to take your kids to a playground where there's a gigantic sign that says: "No Running?" How would you like to take your kids to a playground that has a big antique railroad caboose that says: "Do Not Touch?" How would you like to take your kids to a natural history museum where when they walk in, the "docents" are actually "don't-sense" and they yell at your kid for almost everything they do? What if you could actually see more reptiles at your pet store? The list goes on and on and on.

The moral of the story here is you must know what your business is. Know who your customers are. And most important, drive amazing innovation by knowing what your customers value—all in real time.

The natural evolutionary process of companies is to begin with an entrepreneurial spirit that is highly customer connected. As time marches on, corporations move through their evolutionary cycle, from entrepreneurship to organizational-centric models. This tends to take their gaze away from their customers and toward internal processes.

Companies have tried to deal with this issue by trying to reconnect with the customer through so-called CRM, or customer relationship management systems. Unfortunately, the "McDonald-ization," or data pooling, of customer behavior has become the new focus in terms of observing customer needs. The problem is, it's all wrong. The only way you can successfully identify what your customers care about, and to create great innovations, is to "innovate while walking around."

For an example, at one of the largest electronic retailers in the country, if you go in and try to get some answers to your most basic questions, the sales associates have no idea about the features, the advantages, the operation of the very products that they are selling. Despite the fact that customers are looking for a "consultative sale," that is, they're looking for someone to recommend, and make specific intelligent, educated comments about how and why they should purchase a given technology.

This large electronics retailer hasn't invested anything in the training or competence of their sales staff. But when you go to the register to check out, they'll ask you to go online to fill out a survey on the internet to get a ten-dollar gift certificate. This type of data pooling through typical CRM grabs data that's dead wrong. A ten-year-old could walk onto the sales floor and realize what was wrong. They will never be able to secure meaningful data from these CRM methods that are poorly deployed.

Data pooling has its place, but the ultimate way to invent is to connect with your customer so you know what they actually care about.

6. No Resource Commitment

In order for a product to succeed, the model needs to have *people*, the *product*, and the *plan*. But not surprisingly, innovation also takes *money*. Organizations will oftentimes commit to a technology without providing enough resources for its ultimate success. The result is a lot invested in research and development that never translates into profit for the company because it was short-funded or had insufficient team members and/or time resources to deliver the value to the customer.

Lack of resource commitment, again, is symptomatic of a lack of the right kind of leadership in the organization. When the CFO—typically a cost-cutting champion—drives the business, look out!

7. Bail, Don't Fail

Innovation needs to have several off-ramps throughout the process where team members are comfortable "bailing" from the innovation because it doesn't meet the ultimate goal of delivering customer value.

As you go through the development process, functionality and other types of data emerge that help you determine whether a technology is real. The nomenclature of "failing" has been a very big innovation killer, and in this particular pitfall we should change the vernacular so we can look at bailing as a normal event, a healthy part of a natural innovation process, even rather than saying "fail early."

The term "fail early" has become very popular in the innovation space, but "failing" is not the right term. We need to learn to look at bailing from a project or product that isn't adding value for the customer as a learning experience, and we need to learn that "bailing early" isn't "failing early."

8. "Open" Innovation That Isn't

"Open innovation" is a paradigm that challenges the traditional closed innovation model, which relies on *internal* research and development processes to generate new ideas, technologies, and products. Coined by Henry Chesbrough in 2003, open innovation proposes that firms should leverage *external* sources of ideas, technologies, and market intelligence as well as internal resources to advance their innovation efforts.

Open innovation, in the hands of the right organizations, can provide some real organizational value. But pseudo "open innovation" has become the latest innovation tool—the latest panacea—to save the day. The challenge is that *any* method of innovation—including open innovation—must be systemic. Chanting "open innovation" at your meetings will not create an open organization! At its core, open innovation emphasizes collaboration, knowledge sharing, and networking with external partners such as customers, suppliers, universities, research institutions, and even competitors.

9. Innovation Appropriation

Many organizations, especially research organizations and university organizations, are innovation appropriators. What do we mean by that? Oftentimes, researchers are not rewarded for the successful commercialization of technology, and because many of these organizations are process centered, they're also not punished for—or even measured for—the lack of results.

What are many of these organizations doing? The answer's simple: nothing. If you're not rewarded for commercialization success, and you're not held accountable for process orientation, as is oftentimes the case in large corporations and universities, nothing happens. You file patents, but patents aren't in and of themselves a measure of innovation success.

Instead, you want "innovation championing." You create innovation teams that are rewarded for innovation success by taking smart risks. You also make them accountable for processes that get in the way, that is, hinder the innovation process.

Ask yourself: Is your organization an innovation appropriator or an innovation champion? Winning companies are always innovation champions.

10. Lack of Systemic Innovation

The only way you can become a true innovator is to make innovation a holistic part of your organization, from top to bottom. Using innovation in your finance department to streamline your customers' ability to pay and to give them easy and convenient ways to access accounting information, and creating front-end customer interface systems to make the experience of working with your business truly fast, friendly, and effective—that's innovative.

The fact is that great speakers and writers have talked about the process of creating value in the organization—but this process is only about 10 percent of product innovation. The overwhelming majority of organizational value is in the area of innovation *outside of product innovation.* Systemic innovation rules the day.

It's Not Always About Money

One question that's interesting to ponder is: "What percent of business failures are directly or indirectly related to a lack of innovation? In other words, how often do businesses lose their way strictly out of their inability to effectively innovate?"

Innovation isn't just the ability to solve problems; it's the ability to use a variety of tools to make certain that you've done the right things to earn the money you need to run your business. The overwhelming majority of business failures are associated with the inability to deliver meaningful net value.

Such failures can take many forms, each with symptoms that may be obvious or may be more subtle to most until it's too late. Often, these patterns of failure are observed by members of an organization who, unfortunately, have little power or influence to change them. It really takes a leader—a chaotic changemaker—to recognize the symptoms and put an organization back on course. In some organizations, the behavior patterns are so ingrained that it takes outsiders to change them, if they can be changed at all.

When Process Takes Over: What's Wrong with Innovation Management Systems?

It's important to switch gears a bit to consider innovation as a process. Too many companies focus on the process of innovation, not the result—customer net value. But particularly in big organizations, like any other initiative, innovation cannot be done without at least some degree or form of structure. That said, process actually serves to get in the way and to ingrain old, bad habits in the organization and in the culture.

It's important to take a closer look at process, the "lexicon" of process, and the various ways well-intentioned people put process into play, and how it stifles the innovation process. In analyzing innovation failure, excessive focus on process—internal focus—kills innovation over and over, oftentimes without the corporate victim having any real sense what's happening to them. With process, of course, come systems—the dozens of tools ranging from spreadsheets to elaborate custom innovation management software—used to manage innovation and new product development.

Process is such an important topic in understanding the root causes of innovation failure that it warrants its own chapter.

Take Action!

✗ Experts, pundits, and journalists have long thought about what makes innovation fail; there are lists upon lists of reasons. Many of these lists sidestep the fact that innovation failures tend to boil down to cultural barriers and a failure to focus on customers and net customer value.

✗ There is a tendency to look at the symptoms of failure and "blamestorm" the causes—again leading to lists, and again failing to focus on culture and customer.

✗ Cultural failures include excessive focus on process and ineffective leadership. These root causes lead to a number of other failure "causes," which really turn out to be symptoms: excessive risk aversion, underfunding, too much focus on products and not customers, among many. Good innovation managers address the culture and provide the kind of leadership necessary to move forward, not look backward.

CHAPTER 14

Systems = Good, Chaos = Bad

Everyone in business knows the value of systems.

A system is a repeatable process in your business that can theoretically take place without the direct action of a leader or manager. A system is a method of doing something that can be done the same way, over and over, as efficiently as possible. It allows leaders to focus on future growth and moving beyond mere survival to true prosperity.

The alternative to a well-constructed system is plain old chaos. This is the painful condition known to both start-ups and big corporations, where every day you need to reinvent the wheel. Instead of coaxing the spark of innovation to life, you're running around putting out damaging fires. You can't plan for the future because you're too busy managing the present and its many small crises.

And when you plunge into an era of external chaotic change, in which you're facing unprecedented disruption, not having a sturdy set of systems in place can be *deadly*.

Systems can be simple or complex. A simple example in a small business might be an email autoresponder sequence that nurtures a relationship between you and the people who subscribe to your mailing list. It might be a system that triggers an invoice when a certain part of a project is marked as complete.

The bigger the business, the more systems it will have. Big companies have systems for supply chain, sales, production, hiring, branding, marketing—just about every facet of their operations.

Companies that are built on a franchise model are nothing *but* systems. If you're just counting the number of franchisees and stores, the reigning king of the franchise world is 7-Eleven. With 78,400 stores operating in 20 countries worldwide (as of 2023), 7-Eleven has its systems fine-tuned. If you want to open your own 7-Eleven, pretty much all you have to do is plunk down anywhere from $37,550 to $1.2 million (depending on location and other variables), get the 7-Eleven training, and unlock your front door! Services provided by the home office in Dallas, Texas, include obtaining and bearing the ongoing cost of the land, building, and store equipment; record keeping, bill paying, and payroll services for store operations; and fees and financing for all normal store-operating expenses. The head office for 7-Eleven even pays for the franchisee's water, sewer, gas, and electric utilities.

They've got their systems fine-tuned to a science.

But having massive operational systems does not by itself guarantee success. In a time of chaotic change, your systems must be *flexible and adaptable.* Just ask Blockbuster. At its height in 2004, the video rental franchise giant employed 84,300 people worldwide and had 9,094 stores in total, with more than 4,500 of these in the United States. They ruled the video rental business! But the market vanished, their rigid business systems became obsolete, and in 2010 the company declared bankruptcy.

Some companies—but not nearly enough!—have systems for innovation. But I'll get to that in a moment.

The Best Systems Are Agile

In many ways, systems are amazingly good.

Without systems, people have to solve the same problem over and over again. Systems ensure consistency, promote thrift, and reduce repetitive tasks that don't add value. They guarantee that when you order a Big Mac in Boise, it's the same Big Mac that you'll get in Baton Rouge or Bangor. It also means that McDonald's can scale up production of Big Macs and accurately predict the profit margin regardless of whether the Big Mac is sold in Topeka or Tokyo.

But systems can easily turn nasty. They can change from being a friend of an organization to being its worst enemy.

They can become entrenched and resistant to progress. When external conditions change—as they are in today's business environment with increasing speed and depth—people can cling to established systems in the false belief that what is "tried and true" will save the day.

Russell Ackoff, one of the pioneers of business systems, warned against organizational silos, sclerosis, and fragmentation. Ackoff defined the systems age as beginning after World War II, during a time of growing global and technological complexity. Organizations would henceforth have to deal with "sets of interacting problems," and the key challenge would be designing systems that would *learn and adapt*. He said, "Experience is not the best teacher; it is not even a good teacher. It is too slow, too imprecise, and too ambiguous." Organizations need to learn and adapt through experimentation, which he said "is faster, more precise, and less ambiguous. We have to design systems which are managed experimentally, as opposed to experientially."

The moral of this story is that while systems are mandatory for any thriving business, these systems must be agile. They must bend, not break. They must be capable of being reinvented as conditions change.

For any of these conditions to be met, a system must be as *simple* as possible. If it's complicated and a diagram of it looks like a plate of spaghetti, it's going to fail.

The User-Friendly Apple Operating System

An analogy I like to use is the personal computer.

Remember those ancient days before everyone had a smartphone or even a home computer?

Back in the Dark Ages of the 1970s, when computers were beginning to be made in sizes smaller than a refrigerator, the biggest obstacle to their use by ordinary people was their complexity. You had to learn to use command-line prompts to drive the operating system. Punch cards ruled, and while desktop calculators did the math, typewriters did the word processing.

Apple made using a computer intuitive. The Mac operating system, with its desktop and bitmapped graphical displays, was far easier to use and required less training and expertise than the ubiquitous DOS systems. The Mac was the first truly popular computer with a graphical user interface, a mouse, and the ability to show you what a printed document would look like before you printed it.

What this meant is that the user could, without any special computer training, easily master the system and be productive. The user didn't have to expend time and energy learning how to perform a task. The best industrial systems are like that—they're as simple and easy to learn as possible.

Having read this far in the book, you know I preach that innovation must also be *sustained over time*. Again, look at Apple. Today the company makes the one computer in the world that is as easy to use as a toaster—the iPhone. The most revolutionary thing about Apple's first iPhone was the seemingly effortless way in which nearly every bit of complexity was hidden behind a display of easy-to-understand icons. The iPhone contained no visible "directory structure." Your music was not in a particular place on your phone, requiring you to hunt for it; you accessed it by launching the music player with one touch.

Complex Management Systems Can Be Deadly

In a successful innovation mandate, you'll find simple systems that deliver exceptional enterprise value. This includes not only R&D and production systems but *management systems*.

This seems like an obvious formulation, but all too often organizations get tangled up in complicated and bureaucratic systems that actually *stifle* innovation.

The classic example of bureaucracy-stifling innovation—with tragic results—is the General Motors ignition switch scandal of the early part of this century. Eventually, GM had to recall nearly thirty million cars worldwide and the company paid compensation for 124 deaths.

The fault had been known to GM for at least a decade prior to the recall being declared. The problem was unintended ignition switch shutoff because a part called the "switch detent plunger," designed to provide enough mechanical resistance to prevent accidental rotation, was insufficient.

According to an email chain from 2005 unearthed by investigators, GM's managers estimated that replacing the key ignition switch component would cost ninety cents per car but only save ten to fifteen cents on warranty costs. Somewhere deep in the bowels of the vast GM bureaucracy, the fix was repeatedly rejected until 2006—but millions of earlier cars weren't recalled.

Remember—innovation is *not* narrowly defined as "a new invention that no one has ever seen before." That's much too limiting. Innovation includes *identifying problems and fixing them*. It means *making a change to the status quo to add value to the product or service*.

Four Obstacles to Innovation—and the Solutions

Systems can be beneficial or they can hold you back. They can be written down in company manuals, like the comprehensive systems used by 7-Eleven, or they can reside in the guts of the company culture. The latter variety often takes the form of institutional knowledge, or to put it in familiar terms, "That's the way we do it around here."

If "the way we do it around here" is supportive of innovation, that's good.

If it means being stuck in a rut, that's bad.

A system for innovation must be capable of taking the spark of a new idea and developing it into a source of energy. There are at least four significant reasons why corporate innovation is so difficult—but for every problem there's a solution.

1. Previous Success

Problem: When you sell a product and it does well, you've now set the bar. You've hit goals that you want to exceed in the future. This can create a mentality of "if it ain't broke, don't fix it." The organization learns and codifies what made it successful, which locks in a way of doing business and a set of expectations about current and future success.

Solution: It's understandable that it can be difficult to tinker with a successful product, which is why many leading innovators like 3M aim to generate a set amount of revenues from *new* products. In addition, as we can see with many innovative manufacturers like Toyota and, more recently, just about every other car manufacturer, massive innovations happen in the production *process*, out of sight of the consumer. While Toyotas change very little in their exterior appearance from year to year, improvements are always being made under the hood.

If you have a successful product that people love, you're well advised to carefully preserve your market and your brand appeal. But there are still plenty of ways you should be innovating behind the scenes!

2. Catering to the Existing Customer

Problem: Everyone knows that it costs more to acquire a new customer as opposed to keeping an existing one. You know what existing customers want and it's easy to give it to them.

For an entrepreneur, every consumer is a prospect, and there's no infrastructure or product portfolio to support or defend. But established firms, having achieved sales success and having built a product portfolio, want to lock in their customers. They're more inclined to defend their existing customer base rather than innovate to offer new solutions to new customers.

Solution: The innovative organization knows that consumer tastes change—often dramatically! Consider the soft drink industry. You'd think that Coca-Cola was a foolproof product with decades of customer loyalty. Nope! Sales of the company's iconic soft drinks have been sagging for over twenty years as consumers seek healthier beverages. Sandy Douglas, the company's top North America executive, told trade publication *Beverage Digest* that Coca-Cola is "moving at the speed of the consumer" in its flagship market by evolving both its business model and how it measures success. Shifts in the consumer landscape have inspired the beverage giant to rethink its core success metrics. "We will measure ourselves on what people are willing to pay for our products, not the gallons they purchase," said Douglas. "If you follow your consumer, you're likely to have a good day."

3. Resource Allocation and Project Prioritization

Problem: In any organization, there's only so much money, time, and resources to go around. Corporate innovators often find themselves fighting for limited funds, since the vast majority of resources and dollars are going to support existing products—the cash cows. In addition, executives often can't decide between innovation projects. This leads to half-hearted initiatives and piecemeal innovation that is either ineffective or doomed to fail.

Solution: This is where leaders need to step up and define the culture of the organization. It's incredibly foolish to expect that your product or service will be the same in five or ten years as it is today. To meet the challenge of inevitable massive change, leaders need to mandate an *investment in innovation* with the understanding that new ideas are the lifeblood of the business and are worth paying for. Leaders need to look ahead and take the necessary steps to not only cope with change but to leverage it.

Some industries, such as pharmaceuticals and entertainment, know their products are destined to lose value over time and that creating new products—proprietary drugs, movies, popular music—is the only way to survive. They know they must *innovate or die*. It's a liberating feeling!

4. Leaders and Employees Are Stubborn

Problem: People can be rigid and set in their ways! Both leaders and employees can have a fear of failure or simply an aversion to changing how they do their work. They get set in their ways, and view innovation as a painful intrusion into their comfortable routine.

Solution: Do you know what's *really* painful? Going out of business because of a failure to keep pace with the inexorable changes that happen in every marketplace. That's painful for everyone.

To create a strong innovation mandate, leaders need to be proactive about connecting with their stakeholders. Employees need to fully understand what innovation means and how it's going to be managed in their organization. When there are gaps in that information, employees get nervous and rumors start to spread. Leaders need to clarify gray areas and make sure misinformation isn't spreading.

Call a team meeting and explain what's going on in a clear and concise manner. If your company is big, train your managers to do it. Imagine you're pitching your idea to potential investors—after all, your employees are being asked to invest their time and even their hopes—and start at the beginning. Discuss why innovation is important, and why you're so excited about it. Employees who are afraid to think innovatively won't. Traditional employee training and development often do not include support for idea generation and encouragement to think differently. If you want your company to be truly innovative, then put in place the environment that allows your top managers to teach innovative thinking to their people. If you want to foster an environment of idea generation, then you need to encourage new and risky ideas to be voiced.

Make your innovation operating system no different from any other system in your organization. Fund it properly, educate your stakeholders, and make it simple. Set attainable goals and celebrate both successes and failures. Bring the spark of innovation to every part of your organization and see the powerful results.

Lessons from the World's Biggest Nursing Hackathon

The American Nursing Association (ANA) does an amazing job of serving their members by helping them leverage the skill sets and insights that will affect the way in which they deliver safe and efficacious care. A great example of their embrace of the spark of innovation was the world's biggest nursing hackathon, which the ANA hosted at their annual convention.

In this case, eight hundred participants used innovative thinking to determine ways to advance safe patient handling and mobility, prevent violence against nurses, strengthen moral resilience and ethical practice, and protect health-care workers against needlestick and sharps injuries. Nurses initially generated ideas on their own at group tables, and in a succession of votes, whittled them down to winning solutions.

This ANA hackathon was no small matter, as historically innovation has too often lived only in the corner offices of hospitals and clinics. The ANA recognizes that every single day, nurses get unfiltered, first-hand knowledge of problems and opportunities. On both a short-term and long-term basis, they work closely with patients. (In other industries, they're called customers.) Because of their frontline experience, practicing nurses have more ideas on how to make things better for the nurse, doctor, patient, and organization than anyone else.

During the hackathon, these amazing nurses unleashed their innate power to innovate, solve problems, and identify new opportunities. "We need to get people to believe in their own creativity," said Karen Tilstra, PhD, cofounder of the Florida Hospital Innovation Lab, where nurses and others can bring their challenges and innovate solutions. "Innovation is always a step in the dark," she added. "It takes courage. But you don't have to know everything to start finding solutions."

As the ANA reported, here are just a few examples of nurses' innovative thinking from the Orlando hackathon:

- A relaxing virtual reality room where nurses can take a break from their unit.
- An app in which nurses could report any violent incidents, as well as track the total number of incidents in twenty-four hours.
- Gloves that serve as armor against needlesticks and sharps injuries.
- Are any of these ideas seismic or disruptive? Probably not.

Could these incremental ideas (and others), when accumulated and applied consistently day after day, make a huge difference to an organization's ability to deliver value to its customers and drive up profits? *Absolutely yes!*

The hackathon was an amazing example of opening the floodgates of ideas to get new perspectives from the very individuals who have the best actionable insights to drive innovation. We are now seeing this process ramp up as more and more organizations are beginning to see that collaborative organizations that build out simple but powerful innovation pipelines are constantly leading their markets and new innovations in customer satisfaction.

Take Action!

✗ **Start planning your innovation operating system.** Your innovation mandate begins with the goal of achieving strategic advantage in the marketplace, so in the planning phase you should think specifically about how innovation is going to add value to your strategic intents, and focus on the areas where innovation has the greatest potential to provide strategic advantage. In the well-managed innovation effort, you expect insights to come about as the result of carefully constructed and managed processes and activities, not by random chance.

✗ **Keep it simple!** Complexity does two things: It discourages the sometimes sensitive human beings, who find they must either play politics or thread their way through a bewildering environment to promote a new idea; and it makes even routine incremental advances much more difficult.

✗ **Engage your stakeholders from the top down.**
Innovation will be driven by the people who work in
your organization, and they need to be 100 percent on
board. At the end of the day, the innovation mandate is
all about *trust* and *awareness.* Your stakeholders need
to trust that their efforts will not go unnoticed and
that failure is to be expected, and they need to be aware
that innovation needs to be as natural as any other job
function.

✗ **Position your company for innovation success.**
Like Coca-Cola, align your organization around the
dictates of the market. Empower each team member
to make decisions that apply to their own groups
and roles. For the highest level of engagement, help
your employees align their own self-interest with the
organization's interests. Allow your team members
to migrate into new groups and to align themselves
with their own self-interests within the company.
Hire only people who will think outside the box and
devise unique solutions to complex problems.

PART III

Chaotic Change Is People Powered

Human Experience Innovation™ (HXI) is a powerful new discipline of work that acknowledges the fact that enterprise success is based on an organization's ability to create significantly better experiences throughout its entire experiential ecosystem. The experiential ecosystem is comprised of individuals and organizations that they impact in any way. This includes customers, employees, vendors, partners, and their community.

It's virtually impossible to deliver optimized customer experiences with unhappy employees, and it's equally impossible to have happy employees when you have unhappy customers. It's also true that to create collaborative relationships with others in your ecosystem, the relationship must be exceptional.

If you want the best possible value and components that you procure for your products and technologies, needless to say, you need a great human relationship that's beneficial with your vendors. If you want to attract great local talent, then of course you need to have a great community reputation and relationship. Even in a time of disruptive AI, an organization's success is based on the quality of their relationships.

All of their relationships!

CHAPTER 15

The Five Customer Touchpoints

Across the vast global marketplace of goods and services, the number of relevant customer touchpoints varies. More is not always better! One major international luxury car manufacturer developed a plan with 632 customer touchpoints! Can you imagine how impossible, moreover ineffective, it would be to try to implement a marketing plan with that many touchpoints?

Let's strive to keep things simple, digestible, and actionable. You can usually break down the customer journey into five basic, interconnected touchpoints. They may not be profound, but they're real.

1. The Pre-Touch Moment

This is the research phase, done digitally and nondigitally. The digital pre-touchpoint is where your potential customers are checking you out online. They Google you and/or find reviews on Yelp, Amazon, or some other site. In other words, they're educating themselves about your online reputation before they actually engage with you. Underestimating the value of this touchpoint is asking for failure. Its importance will only grow over the years.

The pre-touch can also be nondigital. If you have an actual physical location, what does it look like to potential customers driving by? When customers park their cars? How about when they walk up and open the door? Are they greeted with smiles or indifference? It all adds up.

2. The First-Touch Moment

As mom always said, "First impressions last a lifetime." So it goes with the first touch. It sets the theme for how your customer will forever perceive your product, brand, or service. If you have a bad first touch, it's really hard to fix. Conversely, if your first touch rocks, then you can easily build on that.

Customers checked you out in the pre-touch; here, for the first time, they actually engage with you. As with all the touchpoints, the key is to build an exceptional experience across all your customer types. Identify those types, understand what they want, then deliver it to them—exceptionally.

3. The Core-Touch Moment

Your customer has bought your product, brand, or service and is now living with it. Your product needs to deliver the goods, day in and day out.

Just gaining customers isn't enough: You have to keep reinventing great ways to serve all customer types throughout the entire relationship.

4. The Last-Touch Moment

The last touch of a particular experience is the final moment the customer has with a product or service. (Actually, there should never be a last touch, because ideally you will have many more touches with all your customers.) At this touchpoint, you send your customer off with a memorable goodbye that makes them want to come back. It's a way to thank them for being with you, to tell them that you valued the experience and hope they did too.

5. The In-Touch Moment

This is how and when you stay connected with your customers after their experience with you has ended. You must approach this with an absolute commitment not to try to sell them anything, but rather to consistently and pleasantly provide them with ongoing value. You want them to willingly come back to you of their own accord, not because you're shoving some one-time-only shenanigan down their throat.

CRM software packages are often used to sell people something. They enable companies to stay in touch for the purpose of getting customers to buy something else. But this isn't what customers love; it's what they hate. Instead, deliver ongoing constant value. Let your customers know about special offers only if you know it's something they need to have.

Take Action!

✗ Science backs up this new body of knowledge about customer service. The only commitment you have to make is a willingness to lean into the information and to accept the discomfort of learning new requirements and a new way of viewing the customer service landscape.

✗ Albert Einstein said, "The measure of intelligence is the ability to change." This certainly holds true for the shifts in power we see today in customer service.

CHAPTER 16

Know Your Customer Types

To identify your customer types, you can't take a set of generic customer types, slap them onto your consumer experience strategy, and then expect to provide your customers with relevant experiences. You must begin at the beginning with your own customers.

To illustrate how to go about identifying customer types and designing relevant, exceptional consumer experiences across various touchpoints, using both digital and nondigital channels, let's work with a fictional company in the luxury car wash and detailing segment. The company is called NeoWash. It's a useful example because its core business is highly traditional and *not* typically associated with chaotic change. It's like any other service company—a hotel, a restaurant, an amusement park, an NFL team. Remember, advances in innovation and disruption are often seen *behind* the office door, in the company's back-end operations where profit margins are made or lost. When you're selling a commodity, like car cleaning, in order to stay one step ahead of your competitors you *must* find efficiencies and innovations in how you structure and operate your business.

The process begins with three simple steps.

Step 1: Brainstorming Session

"Ideation" is a ten-dollar word for brainstorming, the process of forming images or ideas. It's a good place to begin. Start by asking yourself what commonly known attributes of the human experience might impact the customers of NeoWash. For example, you can anticipate that in a car wash, some people are primarily interested in fast service, others in high quality, and still others in low cost above all other traits.

Once you have come up with a range of theoretical customer types, you must test and refine these types using two things:

1. **Digital analytics**—tools used to assess qualitative and quantitative online data about your current and potential customers.
2. **Contact point innovation**—inventing at the point at which the experience is being delivered rather than in a boardroom or laboratory, far away from where customers actually experience the service or product.

In other words, bring your team to the car wash's parking lot and start by seeing how the cars enter and exit.

There was a wise architect who built amazing buildings and then planted grass all around them so that there were no obvious paths leading to the entrances and exits. Weeks later, he simply created concrete pathways on the mashed-down trails that had been organically made by people walking to the doors. This is contact point innovation.

You then refine these types as they engage in the five contact points across your company. The net benefit is that the new experiences you invent across each contact point and customer type will be relevant across your entire range of types.

- If your customers are transactional and all they want to do is get in and buy what they came for, then make that experience as amazing as possible.

- If your customers are experiential, and they want to fiddle with the new gadgets and gawk at the amazing displays, then provide that for them.
- If price is their issue, show them what's on sale or how to save money by downsizing the product.

Different customer types want different things. Try your best to fulfill their expectations. Avoid the easy temptation to stereotype, because anybody could be any type. If the loves and hates are shared, one type could include a thirty-something Latina, a twenty-something rapper, and a seventy-something retired schoolteacher. If you design an exceptional and relevant experience, each of us will respond to it, although our outward demographics are very different.

Step 2: Listening Posts and Contact Point Innovation

The next step is to refine your customer types through listening posts and contact point innovation, both critical in designing the overall customer journey. A listening post literally means having someone listening to your customers. This can be accomplished in different ways—for example, physically, as in standing in line with them, or digitally.

Contact point innovation helps you clarify and refine your customer types, and begin to design exceptional human experiences for these customer types across the range of five contact points via digital and nondigital channels.

Contact point innovation means experiencing what your customers experience at each of the five consumer contact points. You do this by setting up listening posts at each point through both digital and nondigital channels.

At NeoWash, you could begin by experiencing the pre-touch contact point, which could be either digital or nondigital. This is the very first time the potential customer has contact with your business.

Digitally, you might Google "best car washes" in your town and see what came up. Let's say Google shows seven ratings for NeoWash, with an average of 3 out of a possible 5. In the online rating world, this is bad news. You might then go over to Yelp and see five ratings with an average of 2 out of 5. Confirmation of bad news! Then you could read the comments to find out what people hate about NeoWash. Go to other bulletin boards and blogs, as many as you can find, to learn why people think the car wash sucks. Most important, by doing this, you are experiencing firsthand the digital pre-touch contact point.

Not everyone's first contact with NeoWash will be digital. A person who's old-fashioned and "offline" might drive past your car wash and look at it. This is the first nondigital contact. What does your car wash *look* like? What first visual impression does it give? Is there a long line of cars waiting to be washed? Are the windows clean? Is the sign clear and does it look new? Are your staff wearing uniforms, or at least branded work shirts and caps?

The important takeaway here is that these first impressions are critical, and once you know what they are, you can invent new and amazing first impressions across the range of your customer types.

Step 3: Repeat the Process: Undergo the Entire Customer Experience

Don't stop at the pre-touch experience. Each of the five touchpoints needs to be assessed from the perspective of a range of customer types. Remember, the disruptive innovator looks at the universe from the viewpoint of what she can create. She's designing new ways to create exceptional human experiences. She's going to invent ways to solve each broken touchpoint.

Therefore, your next step at NeoWash could be to take a traditional approach and physically sit in the car wash, watching, listening, and asking questions. What sort of people are coming in? What are they experiencing? By doing this, you refine what your customer types love and hate.

You might also identify new types or combine types. You will be able to identify broken systems and opportunities to improve them by inventing human experiences that are more relevant for each customer type. For example, you might learn that church gets out at 11 a.m., and that's why people show up at 11:15. Talk to your customers. Ask them where they're coming from and where they're going. Learn from them.

STEP 1   STEP 2   STEP 3

You also want to experience your employees' perspectives, which as we've seen is critical to excellent customer service. What do they need that they don't have? What's their day like? What do they love and hate about their work? When and how do they interact with the customers? What can you see about your employees through your disruptive innovator eyes? Employees want to deliver exceptional service, but maybe there are policies or systems in place that don't allow them to do this. Maybe they're lacking the resources that would help them get their jobs done better. Listen to them. Talk to them. See firsthand the quality they can deliver.

Through this process, you will begin to understand and dissect your customers and therefore customer types by what they love and hate about both their digital and nondigital journeys. You will then be able to identify ways to add layered value across the five touchpoints to the entire range of types. Remember that when you drill down on any individual customer type, you probably create layered value across all the other types. For example, if you increase the speed of service at the car wash by using time-motion data, it would obviously please customers whose prime interest is speed. But all your customer types benefit from faster service. And when you increase the quality of the car wash for the customers who are primarily looking for that, all your customer types likewise benefit.

Identifying customer types increases your ability to gain the insights necessary to drive disruptive and breakthrough innovations. If you collapse the value of one customer type while inventing for another, then you're not finished inventing. All innovations should have a neutral or positive impact across all the other customer types.

Let's say your brainstorming session, listening posts, and contact point innovation help you find four distinct customer types for NeoWash: Sparkly, Speedy, Thrifty, and Touching.

Type 1. Sparkly

Let's call the first customer type Sparkly. This is a group of people who actually don't care what they pay. They don't even mind how long they have to wait. Their primary love is leaving with a perfectly clean car. They want quality; they hate having to wipe away missed spots and smudges. They're the ones holding up the line, pointing out imperfections to the final detailing guy.

As a chaotic changemaker, you realize the quality control system at your car wash is broken. You are going to fix that broken system to please the Sparkly customer type. You are going to do something totally nondigital: *You're going to hire a senior citizen.* What? Yes! You put him in a bright orange quality control vest, and his only job is to do the final inspection of each car before it's released to the customer. He will be the final quality assurance person. Maybe you even put candy canes in his pocket for the kids.

Designing human experiences is a phrase you'll see throughout this book. When you identify customer types, you design these experiences better, because you're identifying not just what people are, but more important, who they are and how they wish to experience a range of touchpoints across digital and nondigital channels.

Type 2. Speedy

The second customer type is Speedy. He or she has one priority—speed, obviously. They hate waiting; they love fast service. They're busy and digitally savvy. Some days they have a good experience at the car wash, but on other days they must wait far too long and leave dissatisfied. As you identify this customer type, you also identify another broken system: reliably fast service.

Throughput analysis, or fast-track methodology, has been in use for years. You can apply this same methodology to NeoWash. You can identify time inefficiencies throughout the entire process—from onboarding a car through delivering it spotless and clean at the other end. If you get even more innovative, you could hire a NASCAR pit crew to give you their insights, and you could develop an app that would allow Speedy to book their time slot in advance.

This would completely change Speedy's relevant experience. With a reserved time slot, when they arrive they'll drive past the other cars straight to the "Fast Track" lane, which has been designed to look like a NASCAR pit stop. Speedy now knows exactly when their car will be washed, and because it is prescheduled, it will all be done quickly and reliably.

As a by-product, because you used fast-track methodologies and time-motion studies to significantly increase the throughput of cars, you are able to reduce labor costs. So while you were providing a relevant experience to one customer type (Speedy), you are now able to offer lower-cost car washes to other customer types (those who are willing to wait). Chances are all your customer types will appreciate fast service at a fair price. Remember stacking values? That's exactly what this is.

Type 3. Thrifty

Not surprisingly, one of your customer types is primarily cost conscious. These customers' loves and hates revolve around price, so we'll call them

Thrifty. For this type, do a price sensitivity analysis. Talk to customers, look at your competitors' prices, and see what you can do to become more relevant to your Thrifty customer type.

Part of the magic is that as you helped Speedy, you might be able to help Thrifty, who might also want a faster car wash. This is the beauty of stacked innovation. As you identify ways to deliver value and relevancy to one customer, you create a by-product that allows you to deliver benefits to another customer.

In the process of using fast-track methodologies and time-motion efficiency analysis to help out Speedy, you reduce labor costs by increasing efficiencies. This allows you to deliver the most competitive price package to Thrifty. Undoubtedly, other customer types will appreciate this as well.

You could also look at the days and times with the slowest business and offer Thrifty a five-dollar discount if a wash is reserved for those times. This is a good business move for the company and great value for a type who's looking for cheaper car washes.

Type 4. Touching

The final customer type is Touching. This is the person who sees going to the car wash as an experience. It could be the stay-at-home mom or dad or a retiree wanting to get out of the house. For them, it's about the touch, the experience, and the emotive qualities. What does the car wash smell like? Sound like? Feel like? Is there a friendly person there to greet them and a place for kids to play? Is there a nice merchandise area?

As a disruptive innovator designing human experiences for Touching, you invent a sensory emotion across each contact point. You get their shoes polished and have espresso coffee and fresh croissants available in the morning. Touching loves the actual experience, so give them an exceptional one. They may not always have the time to linger, but it'll be an option.

You might even create a Clean Concierge Club. For club members who get their car washed twice a month, the espresso and croissant are free, and there is a separate area with cushy leather couches just for them.

They can sign up for the club online and can accumulate points to be used on extra services.

Whenever a company identifies its most profitable customers and places most of its resources there, these customers are the most profitable because they're subscribing to what that company is doling out.

How about making nonprofitable customers profitable? Understand your customers across all your customer types. Walk around, be disruptive, and look at what you can create that's new and relevant by looking at what you can destroy. This is the best way to come up with disruptive ideas and turn nonprofitable customers into profitable ones.

Take Action!

- ✗ Remember that chaotic change doesn't just happen in "sexy" digital companies. It's happening *everywhere*, even in industries that deal in everyday commodities. These companies fight tooth and nail for every last penny, and leaders who are chaotic changemakers are the ones who consistently stay one step ahead.

- ✗ Do you think you'd find these layers of value if you used only traditional market segmentation? No, you wouldn't. Identifying customer types and delivering relevant human experiences across their entire range creates exceptional, layered value. Do this throughout each contact point and via digital and nondigital channels, and you truly will be designing exceptional and relevant human experiences.

- ✗ To identify and understand your customer types, you must experience what your customer experiences, identify what's broken, and invent relevant new experiences both digitally and nondigitally. Remember to navigate where your customer is across the customer journey. Observe what your customer loves and hates.

Design better experiences across all your customer types.
Execute the innovations that your customer loves.

✗ You're not going to die if you get this wrong the first
time. You can't even break it. What you will do is start
the process of inventing an amazing range of new
value you can deliver. It's a new way of looking at your
customer, which creates the seed of thought for driving
the next big idea.

✗ It takes courage to look at what you're currently
doing and say it's bad. It takes courage to turn what
you're doing on its head and try something else. But
the downside is to do the same old thing and get your
butt whipped by the disruptive innovator down the
road.

CHAPTER 17

The Market Analysis

The fundamental tool of external lucidity is the market analysis.

If your company is B2C, the number one question you must answer is, "Do we have something the market wants and will pay for—even if our customer can't articulate what they want?"

If your company is B2B, the question is slightly different: "How do we find out *exactly* what our potential customer needs, and can we provide it at a profit?"

Either way, your product or service must be in perfect alignment with what your customer needs or wants. This may seem like a no-brainer, but it's a big reason why many companies fail, especially in the era of chaotic change. There are two primary reasons why your product or service might not sell:

1. **The market is saturated.** While some "first adapter" consumers will try any new product just because it's new, they will not provide the growth you need to survive. You need to either sell into a small but exclusive market—think supercars or megayachts—or you need to sell in volume to a broad market. Either way, your product needs to differentiate itself from the

competition. This requires that you know the competition. You have to know their products, prices, level of quality, and availability to consumers, and you have to see an opening where your product fills a gap.

Sometimes physical proximity can be enough of a differentiator, especially for restaurants. For example, in the United States in 2022, the Subway restaurant chain had 21,000 locations. Worldwide, it boasted 37,000 locations. You might think that the last thing America needs is another Subway store! Yet if a neighborhood lacks a Subway location, and the nearest one is far enough away to not be a direct competitor, you can be sure a franchisee will open one there.

2. **The market doesn't need the product.** A classic example of an intriguing innovation that found no market was the ill-fated Segway PT. At its public unveiling in 2001, inventor Dean Kamen promised the Segway—the two-wheeled self-balancing vehicle that used computerized gyroscopes to allow standing riders to travel by shifting their body weight—"will be to the car what the car was to the horse and buggy." Within a year, he projected the Segway factory would be cranking out 10,000 machines a week and make $12 billion in sales. That never happened. Why not? The machine was too expensive, too heavy, and its batteries limited its range. It was also *dangerous*. In 2010, UK entrepreneur Jimi Heselden—who'd just bought the troubled Segway Corporation—was killed on his Yorkshire estate when he drove his Segway off a cliff.

In 2015, Segway was acquired by Ninebot, Inc., a Beijing-based transportation robotics start-up. Five years later, Ninebot ceased production of the Segway PT and laid off the 21 employees working at the Bedford, New Hampshire, plant. Only 140,000 units were sold during the lifetime of the product, and in the later years the Segway PT made up only 1.5 percent of total company profit. Today, Segway-Ninebot manufactures and sells a wide variety of conventional electric personal vehicles.

A market analysis is a thorough assessment of a market within a specific industry. It includes the dynamics of the market and growth potential, volume and value, potential customer segments, buying patterns, competition, and other important factors. A thorough marketing analysis should answer the following questions:

Who are my potential customers?

How many potential customers are there?

Is my product in alignment with what they want?

How much are customers willing to pay for my product?

What are my customers' buying habits?

Who are my main competitors?

What are their strengths and weaknesses?

The Funhouse Mirror World of Market Research

Leaders who fancy themselves to be visionaries like Steve Jobs often cite his comments on his lack of use of market research. Jobs famously said to *Fortune* magazine, "We do no market research. We don't hire consultants. The only consultants I've ever hired in my ten years is one firm to analyze Gateway's retail strategy so I would not make some of the same mistakes they made [when launching Apple's retail stores]. But we never hire consultants, per se. We just want to make great products."

And he told *Businessweek*, "It's really hard to design products by focus groups. A lot of times, people don't know what they want until you show it to them."

These remarks have been misused by reckless entrepreneurs who have embraced the concept of "fail fast"—you design a product, rush it to market, and if it flops, you cancel it quickly.

What's a chaotic changemaker to do? Sometimes the world of market research looks more like you're wandering through the funhouse mirror ride at the theme park. What's real? What's an illusion?

To get back to Apple, in fact they do plenty of market research.

In 1997, Jobs spoke to attendees of the Apple Worldwide Developers Conference where he clearly stated that customers must be placed at the center of operations: "One of the things I've always found is that you've got to start with the customer experience and work backwards to the technology." He then added that the questions Apple asks are, "What incredible benefits can we give to the customer? Where can we take the customer?"

They may not do focus groups, which asks consumers to speculate on the future, but they do consumer surveys. Apple takes great pains to understand the needs of their customers, and surveys them to supplement their own internal data and thinking. This came to light in 2012 when, during a legal conflict with Samsung, the company's VP of product marketing submitted a document to the court explaining why documents relating to Apple's market research (specifically iPhone surveys) should be kept secret. One was an "iPhone Owner Study" labeled "Apple Market Research & Analysis, May 2011." It surveyed users in multiple countries about why they bought an iPhone. (As it turned out, "Trust in the Apple brand" was number one.)

Feedback surveys have proven to be an effective way for Apple to gather customer insights. The company emails surveys to customers immediately after they have made a purchase. Customers are asked to rate their satisfaction level and how likely they are to purchase again.

The Four Key Axioms of Customer Surveys

Just like employee surveys, customer surveys can be very helpful (as we'll see in the pages ahead), but sadly, most of them are unreliable and annoy the customer. They ask irrelevant questions, such as your age. They ask too many questions and make the customer feel like an unpaid employee or a lab rat. They're designed to elicit praise and avoid pain.

The RealRatings approach—developed after years of experience in the trenches with both spunky start-ups and global corporations—is different. It follows the four key axioms of customer surveys and provides the chaotic changemaker with actionable, accurate information.

1. Keep It Simple

Your customer is giving you their time, and their time is valuable.

The number one rule of voluntary customer surveys, where you ask the customer without any forewarning, "Please take our survey," is that *it must be brief.* With just a handful of multiple-choice questions, it must take the customer no longer than two minutes to complete.

If you try to detain them too long, you'll lose them. In an OpinionLab study, 52 percent of respondents indicated they would likely abandon a survey after just three minutes. Even if people make it past three minutes, the quality of their survey responses declines as the respondent hurries through to the end. Keeping your questions brief and simple to understand is the best way to combat this kind of response fatigue.

2. Ask Relevant, Actionable Questions

Because your survey is short, the questions must be important and the answers actionable. If you ask, "Would you buy another product from us?"

the answer means *nothing*. If the customer says "no," what are you going to do? You have no idea *why* the customer is unhappy, so what can you improve? Price? Quality? Service? The smell of the store?

Surveys should be narrow. You cannot hope to learn everything from a survey, so you need to decide exactly what you want to know about. Then after a period of time, you change the survey to address other potential problems.

3. Look for the Hatepoints

Most traditional surveys are designed to elicit positive responses. Customers often try to be polite and tell the survey sponsors what they think the sponsors want to hear. The results are bland mush.

Asking customers what they *hated* about their interaction with your brand can be scary! No one wants to hear bad news. But it's a bedrock principle of lucid leadership. Sure, if compelled to name something they hated about their experience, they may say something silly. But more often than not, you'll get a golden nugget of information that will help your company improve.

Remember the Golden Rule of Retail:

The customer wants exactly what they envision. They want it now. They want it at the lowest possible price.

You need to know in what ways you are *falling short* of meeting this ideal state.

4. Give Your Customer a Reward

When you ask a customer to participate in a survey, you're asking them to provide two things of value—their time and their opinion. You don't give away products for free, so why should your customers give you free stuff? Reward them with a discount, a gift card, entry into a sweepstake, or something else that's easy and fun.

The Next Level:
The RealRatings Customer Survey

Chaotic changemakers often use four customer personas—the Driver, Analytical, Amiable, and Collaborative—to help construct an employee survey. To be truly effective, a survey needs to adapt itself to the *expectations of the consumer*. Of course, you can say that every consumer wants the best possible product at the lowest possible price, delivered ASAP. But the *process* by which the consumer *discovers* and *chooses* the product can vary greatly. This means that from one consumer to another, their expectations of the process can be very different.

The Driver knows what she wants and expects prompt, efficient, no-nonsense service. She isn't looking for friendship and will quickly pile on the hate if her expectations are not met.

The Analytical needs to know every detail of the product or service. Think about the person who goes to a restaurant and before ordering the salmon interrogates the server about its country of origin, whether it were wild caught or sustainably farm raised, and if it had been fed processed meal. That's the Analytical.

The Amiable wants you to approve of their purchase of the product or service. They expect a high level of personal service. An Amiable will spend half an hour chatting on the phone with the Zappos customer service rep before ordering one pair of shoes.

The Collaborator needs the consensus of a group or partner in order to proceed. The collaborator is the person who will say, "I like it, but I need to talk to my partner before we buy."

Of course, there are shades of each—for example, a Collaborator may show some characteristics of the Amiable—but as a general concept, the system hits the mark.

These types each view your company through their own eyes. If asked to describe the sales staff at your company, the Analytical might say, "Their sales team is highly informative. They really know their products." The Driver might say, "The salespeople take care of business quickly and efficiently." The Collaborator appreciates your patience and lack of pressure while he consults with his partner. The Amiable might say your staff is exceptionally friendly and welcoming ("They serve you free coffee and doughnuts.")

And guess what? They're all talking about the same sales staff! It's because customer-facing employees have been trained to adapt their behavior to match or mirror that of the customer. They know how to pivot to meet the expectation of each customer persona.

The Qualifying Question

The first question you ask of your customer will qualify them into one of the four customer types. (Again, this is an example. Your company may need more than four.) For example, for a store called Millie's Fashions, the first four survey questions could be this:

> When I came to Millie's Fashions, I wanted to:
> A. Make my purchase as quickly and efficiently as possible.
> B. Take time to explore the store and get to know its staff.
> C. Learn all I could about the item before I bought it, such as where it was made.
> D. Have the opportunity to consult with my partner.

Because this is just a qualifying round, the respondent can be asked to choose the one that's closest to what they expected. This will ensure just one choice.

Survey for the Amiable

Based on their choice, the respondent is then presented with one of four different surveys. The survey presented to Drivers will be different from the survey presented to Amiables, Analyticals, and Collaborators. The survey presented to Amiables will focus not only on the ubiquitous questions of product quality, but on questions designed to measure to what extent the company scored lovepoints and/or hatepoints based on their specific expectations.

After the initial qualifying question or questions, the respondent is asked to answer two sets of questions. There should be no more than five in each set, for a maximum number of questions set at ten, plus the initial qualifying question.

In the first set of five questions, the Amiable respondent is asked what they loved about the experience. In the second set of five questions, they're asked about what they disliked. We *want them* to tell us something they hated. The love-hate sequence goes like this:

1. **Lovepoints.** Using the scale of 1 (unliked), 2 (liked), 3 (loved), or 4 (really loved), please indicate your level of *happiness* with the following five events:
   A. My interaction with the first salesperson I spoke to when I entered the store.
   B. My interaction with the salesperson who rang up my purchase.
   C. How long I had to wait in line to pay.
   D. Learning about the store's returns policy.
   E. The overall vibe of the store staff. Do they seem like nice people?

2. **Hatepoints.** Using the scale of 1 (not good), 2 (bad), 3 (hated), or 4 (really hated), please indicate your level of *dissatisfaction* with the following five events:
   A. How long I had to wait before a salesperson greeted me.
   B. The time spent looking for the item I wanted.
   C. Having my questions answered by the salesperson.
   D. The checkout process.
   E. The level of attentiveness of the salesperson to me and what I wanted.

As you can see, the questions presented in the two parts are similar but not exactly the same. They are primarily focused on what's important to the Amiable, which is their interaction with the staff. The reason we ask specifically for what the customer may have hated was to give the customer "permission" to express their true feelings. We're asking them what they didn't like about their experience, and we want them to tell us.

Survey for the Driver

In the first set of five questions, the Driver is asked what they loved about the experience based on the criteria most important to them. In the second set of five questions, they're asked about what they hated. The love-hate sequence goes like this:

1. **Lovepoints.** Using the scale of 1 (unliked), 2 (liked), 3 (loved), or 4 (really loved), please indicate your level of *happiness* with the following five events:
    A. The layout of the store and ease of navigation.
    B. My interaction with the first salesperson I spoke to when I entered the store.
    C. How quickly I was able to locate the product I wanted.
    D. How long I had to wait in line to pay.
    E. The overall vibe of the store staff. Is the operation highly efficient?

2. **Hatepoints.** Using the scale of 1 (not good), 2 (bad), 3 (hated), or 4 (really hated), please indicate your level of *dissatisfaction* with the following five events:
    A. The time it took me to buy the product I wanted.
    B. The exact match of what I wanted versus what was available to buy.
    C. Having my questions promptly answered by the salesperson.
    D. The price of my purchase relative to my expectations.
    E. The overall "hassle-free" nature of my experience.

The questions presented in the two parts are similar but not exactly the same. They are focused on what's important to the Driver, which is their ability to get in and out of the store quickly and with the product they want. We ask what the customer hated to give them "permission" to express their true feelings.

Could a single survey be designed that would capture the same information from each of the four customer types? Probably, but it would be exceedingly long. Few customers—and absolutely no Drivers!—would be willing to take the time to complete a long survey. Eleven questions—the qualifier and the two pairs of five each—are about as much as most people will be willing to give you. Anything longer will get you a right rate of noncompletion and careless answers.

Take Action!

- ✗ Know your market! Don't guess. Don't fancy yourself to be another Steve Jobs who can peer into a crystal ball and see the future. (He couldn't, and you can't either.) A smart market analysis will tell you if consumer (B2C) or another business (B2B) would want to pay money for your product or service.

- ✗ You must ensure that you're serving your customers in the way they expect. Make it as easy as possible for them to buy from you! For example, when you want to order a pizza from Domino's, there are 15 different ways (at last count) you can do it. How does your sales system measure up?

- ✗ The lucid leader has his or her finger on the pulse of customer opinion. Your marketing people need to monitor and respond to Amazon Customer Reviews, Angie's List, Trustpilot, TripAdvisor, Yelp, Foursquare, and any other digital platform where your customers might either praise or condemn your company.

✗ The RealRatings approach follows the four key axioms of customer surveys and provides the lucid leader with actionable, accurate information. The four axioms are:

1. Keep it simple.
2. Ask relevant, actionable questions.
3. Look for the hatepoints.
4. Give your customer a reward.

✗ The RealRatings Customer Survey measures how well you are meeting your customers' expectations, with the understanding that different customer types have different expectations, and therefore should not be treated the same.

The Driver knows what they want and expects prompt, efficient, no-nonsense service.

The Analytical needs to know every detail of the product or service.

The Amiable wants to be your friend.

The Collaborator needs the consensus of a group or partner in order to proceed.

The Qualifying Question is critical. Using the same "skip logic" we used in the RealRatings Employee Survey, it guides the customer to the survey questions designed for him or her.

CHAPTER 18

Employee Emphasis Archetypes

Customers and employees have one important feature in common: They both comprise groups of human beings who *want to be happy*. Customers want to be happy buying your product or service, and employees want to be happy working for you. So in a very real sense, they're both your customers because they're engaged in an exchange of value with you, in which your ultimate goal is to earn a profit.

Just as customers fall into various archetypes, so do employees. They are remarkably similar!

In the RealRatings survey system, we talk about employees being categorized by their "emphasis archetype," so named because each highlights an *emphasis* on different aspects of their happiness at work. We're not looking for purity or exclusivity, but emphasis or tendency. There are positive emphasis archetypes and negative ones. Organizations should go through the process of ideation to identify the range and variety of the archetypes working within their organization.

This is a custom process. It's not one-size-fits-all. In all their myriad behaviors, dreams, and expectations, humans defy facile categorization. While it would be easy for an expensive consultant to provide you a range of generic archetypes, tied up with neat little labels, this would be a tre-mendous disservice to you. Each organization, culture, marketplace, and enterprise has its own unique and special environment that deserves the customization of thoughtful personification.

The following are just a few examples of archetypes that you may have in your organization.

The Warrior

The warrior archetype is on a success track. They're all about winning, earning, and growing their career. The warrior archetype is matrix-driven; everything they do is measured, and often the yardsticks are career path-way, income, and the prestige of the title.

The warrior sees their career as a battle they want to win. They're prag-matic, smart, and even aggressive. They're interested in your mission to the extent that you create the right messaging to link your organizational mission to the warrior's personal mission.

There are, of course, good warriors and bad warriors. Good warriors are missiles with an ethical guidance system, whereas bad warriors have no ethical guidance system and often find themselves in deep water. You can't fix a bad warrior, so you should neither hire one nor retain one that you have.

Good warriors are incredibly effective, but they need to be incen-tivized by leveraging the right messaging, career pathway design, and opportunities. Chances are you have warriors on your team today. Do you understand what success looks like to them, and have you created linkage between what's right for them and what's right for your enterprise?

Warriors Love . . .

- Challenges and playing games they can win.
- Moving quickly and having well-defined success targets.
- Having robust resources to achieve their mission.
- Winning!

Warriors Hate . . .

- Playing games that they can't win.
- Obstacles and bureaucracy.
- Being resource-starved or left unsupported.
- Losing!

The Analyst

The analyst is all about the cause-and-effect relationship between what they do and its meaning to them. The analyst is incredibly thoughtful about the linkage between the activities that they do each day, the results they achieve, and a clear and crisp view of a path moving forward.

The analyst is about data. They are typically pragmatic, introspective, and value careful thinking. They ask you, "Why are we doing this?" They pay attention, and they tend to hate inefficiencies and waste. (God bless them for that!)

Just like with warriors, there are good and bad analysts.

The good analyst believes in the message and they are committed to the mission and their job. The good analyst uses their analytical skills to serve the organization and their teams. They look for messaging and systems and tools that create intelligent connections between their reverence for data and facts and the overarching good of the organization. They see analysis as a necessary prelude to *action*. Without action, all the analysis in the world is pointless.

The bad analyst is constantly using their analytical senses to find what is wrong. They're winning at identifying what's not working but are rarely able to offer thoughtful and effective solutions. In fact, they're not really interested in fixing anything. For a bad analyst, "analysis paralysis" is a comfortable place to be. Theory is much more important than action.

Winning leaders don't hire bad analysts, and it's hard to fix a bad analyst—they need to be let go. Winning leaders are eager to hire good analysts, and they incentivize them by being very thoughtful about connecting their analytical minds with the overarching good of the organization, its mission, and the analyst's career pathway.

Analysts Love . . .

- Data.
- Proof.
- The safety and security of data.
- Well-defined processes and structure.

Analysts Hate . . .

- Meaningless theories and speculation.
- Unsubstantiated facts and figures.
- Reckless leaders who operate by instinct, only to go down in flames.
- Amorphous or poorly structured processes and policies.

The Pragmatist

As the name suggests, the pragmatist is down to earth. They are good at performing for leaders who connect with their innate sense of practicality. Winning leaders will give them clear, crisp instructions with goals and measurements, and the pragmatist will take them and run with them.

There are good and bad pragmatists.

The good pragmatist can be adaptive and flexible, and is able to work outside the margins of what's familiar. Conversely, the bad pragmatist is very rigid and resentful of disruption, and as a result they deliver limited value to the organization. Winning leaders don't hire rigid, bad pragmatists, and they don't tolerate them for long if they happen to be grandfathered in from a previous administration. For the good pragmatist, winning leaders create messaging, tools, and resources to support them, and seek to provide an equilibrium between what's good for the pragmatist and their expectations and what's good for the enterprise.

The Pragmatist Loves . . .

- Common sense approaches toward processes and management.
- Doing things that have been proven to work.
- Incremental, controlled change.
- Clear direction from leadership.

The Pragmatist Hates . . .

- Theories unsupported by evidence.
- Taking the risk of trying something unproven.
- Rapid, uncontrolled change.
- Fuzzy direction and poorly described requirements.

There could be thousands of different archetypes, and you know best what your employees love and hate. Remember, this is not about pigeon-holing someone; this is about understanding somebody so well that you can deliver thoughtful, customized, and beautiful experiences for them, and drive unprecedented levels of productivity while concurrently providing beautiful quality of work life.

Take Action!

✗ As a chaotic changemaker, your job is to keep up with your employees and their needs and expectations. This is no different from how you approach your customers. You see how your customer base is changing, and the fact is that your employees are from the same pool as your customers, so your employee base will change as well.

✗ Your employees need to interact with your customers as peers. For example, if you sell Rolls-Royce motorcars, you need employees who are comfortable serving the people who buy high-end luxury cars. If you sell burgers in an inner-city fast-food restaurant, you need—you guessed it—employees who are comfortable in that environment.

CHAPTER 19

The Employee
Touchpoint Journey

The core of the concept is this: At each of the five touch-points, your customers have different expectations and experiences. You need to be aware of this and architect the customer's journey accordingly. From the first touch, which is before the customer has any substantive interaction with your brand or your people, to the last touch, which is after the sale when your customer is once again a "free agent" and may or may not choose to buy from you again, your relationship with your customer is evolving. It requires that you stay focused on the reality of the moment—not what you *hope* will happen or *think* should happen but what's *actually* happening with your customer.

The touchpoint journey with each of your employees is much the same, but in many ways even more volatile and fraught with risk. When you think about it, all customers have one commonality, in the sense that your transaction with them is always the same: They give you money and in return you provide them with your product or service. Some customers will spend more than others and their spending may influence your business strategy. But unless you have a very customer-centric B2B business, your customer is always on the *outside*, while you're on the *inside*. You don't have customers sitting in on your manager's meetings or hovering over the assembly line. Think about a business like a bank. In that case, you actually serve your customers from behind a protective screen, and if they tried to come around behind the counter, you'd call the police.

On the other hand, an employee is someone you bring into the company to work with you every day, shoulder to shoulder. Their job is to help the company reach its goals by either carrying out policy (for employees and managers) or making policy (leaders). In a typical consumer goods or services company, any single employee will have more influence on the success or failure of the enterprise than any single average customer.

This is why the employee touchpoint journey is critically important. You need to understand it and ensure that at every step your employee is happy and therefore productive.

The employee's journey with your company comprises a series of five stations or touchpoints. At each of these touchpoints, the employee and the company build a relationship through countless daily interactions. These interactions may be brief or prolonged. At every touchpoint, the company has the opportunity to create either a feeling of love and happiness by the employee or a feeling of hate.

The sum total of the employee experience over the five touchpoints—which may extend in time to cover many years—is the Net Employee Experience (NEX). This is the measurement of what the employee hated and loved across all five touchpoints.

As we know from the RealRatings system:

Hatepoints are measured across four negative experiences: not good, bad, hated, really hated.

Lovepoints are measured across four positive experiences: unliked, liked, loved, really loved.

Net Employee Experience (NEX) represents the net total of the work experience for an employee, found by subtracting the hatepoints from the lovepoints to produce a score. This score is very useful because it's expressly asking what an employee didn't like and what he or she did like at a specific touchpoint. This provides actionable insights that an organization can use to rapidly fix the dislikes and significantly improve their NEX score.

In the following five sections, we'll discuss the five touchpoints and the risks and opportunities of each as they relate to the overall NEX score.

1. The Pre-Touch

The first touch is not really a touch at all, but the awareness of your business in the mind of the future employee.

During some period in the past, your future employee had no knowledge of your company or your products, and therefore no opinion of it. This general lack of awareness may have lasted for a few years into childhood or well into adult life. What's the first brand name that infants learn about? Probably for most little kids it's the Disney Company—and the company strives mightily to ensure the impression is positive. Remember, most people learn about companies first as consumers and only later as prospective employees.

On the other hand, if your company is a start-up, then most of your prospective employees will have no prior knowledge of you. If in 1998 you asked software engineers and coders if they had heard of an internet company named "Google," the vast majority would have replied, "Google? No." Back then, the business had just two partners—Larry Page and Sergey Brin. The very first employee they hired was Craig Silverstein, a fellow PhD student at Stanford. By 2021, the company had nearly 140,000 employees working in 78 offices in more than 50 countries worldwide.

These days, if you haven't heard of Google, you've probably been living in a cave with no internet.

The pre-touch phase may last for years if your company is a ubiquitous brand name like Ford or Apple. These are names that we learned about as kids, and they've lasted long enough so that they're still relevant (unlike Oldsmobile, Pan Am, Borders Books, and other venerable brands that have vanished). Or the pre-touch phase may last for a very short period of time, between when your future employee sees your job listing on Indeed.com and thinks, "XYZ Company? Never heard of them, but they're looking for someone with my qualifications," and when they walk in for their first interview.

Many companies make efforts to project a positive image to potential employees, especially those like Walmart that must hire thousands of new employees every year. Such campaigns tend to ramp up when the economy is hot and the unemployment rate is low. Then when a recession hits, companies shed jobs, and labor becomes cheap, suddenly it doesn't matter so much!

Much of what prospective employees think about a company is out of its direct control because opinions are formed on the basis of news reports and social media. For example, in the railroad industry in America, companies like Norfolk Southern and Union Pacific have social media reputations that—deserved or not—make them sound like terrible places to work. They can certainly influence their social media brand, and they should, but ultimately the power resides with employees and customers. If you're looking for a career in railroads, their big competitor, BNSF, has a far better reputation, and you'd probably make it your first choice.

Companies can change their reputations. Comcast was once one of the most loathed companies in America, and consumers equated it with the dark and creepy 1996 Jim Carrey comedy *The Cable Guy*. In 2004 and 2007, the American Customer Satisfaction Index (ACSI) survey found that Comcast had the worst customer satisfaction rating of any company or government agency in the country, including the Internal Revenue Service.

But by 2020, a massive internal effort had turned the company around. That year's ACSI survey found Comcast's Xfinity brand was the "most improved" overall, and especially saw a "staggering improvement" in satisfaction with its pay TV service. The employees seem reasonably happy, too—in December 2021, the company scored a respectable 3.9 rating out of 5 on CareerBliss.com, a 3.8 on Glassdoor, and a 3.7 on Indeed.com.

How do you manage your reputation in the marketplace of potential customers, investors, and employees? That could easily be the subject for an entire book, so suffice to say you need to fight the war on two fronts:

1. **Your company must actually be a good place to work.**
   You must strive to instill the happiness principle at every level and in every cubicle and workstation. It must be real and not just a public relations scam. Thanks to social media, companies

are increasingly transparent. What happens in the office quickly becomes public.

In July 2021, a video went viral on TikTok, receiving over 3.7 million views, nearly 550,000 likes, and thousands of comments. Posted by "amazonassociatef1," it was filmed when workers at an Amazon fulfillment center headed for break at 9:59, which was one minute early. A disembodied voice booms out over the loudspeaker, "It's 9:59, not ten o'clock, guys. If you are on your way to break you are getting TOT as we speak!"

"TOT" stands for "time off task," something Amazon will deem an unpaid break, for which the company will not compensate employees.

Episodes such as this one served to draw public attention to the fact that warehouse bathrooms can be located as distant as a 10-minute walk from a worker's station. This makes it nearly impossible for these employees to use the restroom during a 15-minute break.

Even more dehumanizing than the Orwellian tracking was the fact that the supervisor yelled at the employees over a loudspeaker rather than personally speaking with them to find out what was going on.

2. **You need to stay in touch with your potential labor pool.** Aside from actually being a happy place to work, you need to ensure that your story is being told. You need to be on top of social media and know what's being said.

Let's face it—if you have 100 employees, then it's likely that one of them will take a swipe at you on Glassdoor or Indeed. That's just the way life is! It happens to the best of companies. But if a highly qualified prospective employee checks out your company's reviews and sees one or more negative one-star reviews, he or she won't be swayed by them if they see 99 others that are positive, and if your overall rating is high.

Every company of every size needs a social media director. If your company is a sole proprietorship, then that person will be you. You'll need to create your social media presence. This means learning about the various platforms—Facebook, Instagram, Twitter, Snapchat, even TikTok—as well as the employment websites.

As you add employees, you can shift the burden to a professional, full-time social media manager.

2. The Interviews

Let's say you're looking to fill an executive-level position. The person in this leadership job will wield enough responsibility and power to materially affect the current and future performance of the company. For whatever reason, it's not being filled internally, so you need to search outside the company.

The process is neither simple nor cheap.

Your goal is to attract the best possible "happy warrior" you can. In a growing economy, competition for the best executive talent is keen, and you may have to sell your company to the candidate as much as they need to sell themselves to you. Aside from the mechanical stuff like placing ads and hiring headhunters, you need to design an interview process that will give the right candidate a very good feeling about the organization and its direction.

You'll probably form a search committee to identify, screen, and interview prospective employees. This job may also be initiated by your human resources department or in a small company by the person who will manage the newcomer.

It goes without saying that your personal behavior must always be cordial, respectful, and businesslike. And I'm sure you know about the federal and state regulations governing the types of questions you can ask. But in addition, you need to be able to tell the candidate *why* you need them and *what* you want them to do. This will give both parties the opportunity to discuss the specifics of the role and how the candidate can contribute to the success of the organization.

Be sure to leverage the input from the search committee to reach consensus on what you're looking for. To ensure the picture you paint for the candidate is crystal clear, use the MOC framework. Some people call it a candidate profile or job spec, which are the same thing. The goal is to define what you want in these three areas:

Mission: What do you need this person to do over the next year or two? Describe the essence of the job's daily activities in one page or less. Unless the job is technical, like software coding, use plain English, not business mumbo-jumbo. Be sure to specify the amount and type of support and collaboration the new hire can expect to receive.

Outcomes: As the result of the person's work, what well-defined deliverables are you looking for? Ranked by order of importance, list the specific and measurable business outcomes you expect as a result of the mission.

Competencies: In order to execute the mission and deliver the outcomes, what specific experiences, skills, and accomplishments should they have? Which of these are critical attributes and which are optional? These decisions should flow from the mission and outcomes.

Everyone's input should be consolidated into the initial draft of the MOC, and then have the hiring committee meet in person to review the draft.

If possible, have the company CEO personally reach out to the top candidates with a brief, friendly email. This personal connection, which says "you are important to us," can often mean the difference between your top candidate choosing you rather than your competitor.

During the interviews themselves—for high-level positions there will be several—after you've told the candidate what you're looking for, ask them for what *they're* looking for. Many seemingly happy hirings have quickly soured because the interview process failed to reveal the true personality and aspirations of the candidate. Be sure you're not projecting onto the candidate what you *want* them to be, rather than seeing them for who they really are.

Hiring for executive positions can be costly. Expenses you'll be paying may include advertising the opening, the time costs of an internal recruiter, the recruiter's assistant in reviewing résumés and performing other recruitment-related tasks, and the person conducting the interviews, drug screens and background checks, job fair and campus recruiting costs, and various pre-employment assessment tests. If you use an outside recruiter, you can add third-party fees, such as agencies, travel expenses for recruitment, applicant tracking system (ATS) fees, and aptitude test providers.

According to data published by The Hire Talent, a talent assessment company, hiring an entry-level employee is estimated to cost 20 percent of that employee's annual salary. Hiring a mid-level employee typically runs up an average cost of $60,000, or as much as one and a half times the employee's salary. To hire an executive-level employee typically costs a U.S. company more than 200 percent of the new executive's salary to complete the hiring and onboarding process.

At the end of the day, after your candidate has completed the entire interview process, the goal should be to have a candidate who thinks to themselves, "I hope they call me with good news. I really want to work there!"

3. Onboarding and Training

You can interview a candidate over and over again, and get their references, and give them tests, but the ultimate proof lies in just one thing: bringing the person into your company and making them part of your team.

The process cannot be haphazard. You can't just make it up as you go along. Because how you handle the first few days and months of a new employee's experience is crucial to ensuring high retention, onboarding new hires should be a strategic process that lasts at least one year.

Successful onboarding covers four key areas:

1. **Social.** The new employee needs to feel welcome, be able to build and promote productive relationships with colleagues and managers, and become a valued part of the organization.
2. **Operational.** The new employee needs the right tools, materials, and knowledge (such as clarity and business jargon) to do their job well.

3. **Strategic.** The new employee needs to know the mission, vision, and goals of the organization and identify with them.
4. **Ethical.** He or she needs to know the organization's standards of behavior and how they are expected to treat customers, colleagues, and subordinates.

To ensure your new employee is happy, take steps to minimize confusion and uncertainty while maximizing confidence and clarity. Provide an onboarding portal with content designed to engage them, like first-day information, a friendly note from their manager, welcome messages and photos from new teammates, a glossary of company acronyms, a virtual copy of your employee handbook, and information about the nuts and bolts of everyday operations, beginning with where they can park their car and how the phone system works.

Every new hire needs a mentor or guide who's responsible for answering questions and "showing them the ropes." If their immediate supervisor is not able to do it, then it needs to be a fellow team member.

Effective onboarding practices should reflect well on the company's brand. In the first weeks of starting a new job, friends, family, and former colleagues will ask the person about their new employer. You want this first impression to be as positive as possible. With proper onboarding, new employees will immediately share a good impression of the company within their direct network.

That's what you want! Happiness through and through. But despite your best efforts, it can be difficult to know if a new employee is truly happy. Some people will tell you they're perfectly happy right up until the moment they quit. This is because they want to hold onto that power. They may be afraid that if they bring up a problem or say something negative, the company might let them go or somehow penalize them. They want to be able to make that choice, not the company.

Welcome Them—and Then Pay Them to Quit

There are two ways to solve the problem of employees who hide their unhappiness.

The first way is to welcome the new hire to your organizational family. Show them a culture of happiness, self-analysis, and constructive criticism. The new hire will either accept it and become engaged, or will remain suspicious and aloof, hiding their true feelings.

The second way is to make quitting as easy as possible. You can do this by offering them cash to leave.

The practice came to the public's attention in 2004, when Tony Hsieh, founder and CEO of online shoe retailer Zappos, announced that after a week or so on the job, new employees could take a $1,000 payoff to quit. It became known as "The Offer," and when people thought about it, the idea made sense. It's well-known that an unhappy employee can be a financial drain on a company due to low productivity and more frequent mistakes, so it made sense to pay the unhappy employee a small portion of what they cost the company and be rid of them. Soon, The Offer was increased to $4,000.

The Offer also had an effect on employees who chose to stay. Employees who declined the offer were psychologically recommitting to the company. This made them more engaged, more productive, and ultimately boosted the company's bottom line.

As Zappos training manager Rachael Brown told the American Management Association, "The people who do take The Offer are so thankful that we understand that this just isn't the right place for them. But the employees who *don't* take it are very committed. They go home and they come back with these great stories about how they told their family about this crazy offer that Zappos asked them to quit today, and they definitely come back with a sense of commitment."

In 2009, Amazon acquired Zappos, along with The Offer. In his 2014 annual letter to shareholders, Amazon founder Jeff Bezos explained the company had added a program modeled after the one started by Zappos

CEO Tony Hsieh. After some tweaking, Bezos and Amazon titled the program "Pay to Quit." The program worked differently than at Zappos. Rather than receiving the offer once during training, employees at Amazon's fulfillment centers can claim the offer once a year, in February, following the holiday rush. In the employee's first year the offer is for $2,000. The amount then increases by $1,000 each year to a maximum of $5,000, where it remains. The idea is that the longer an unhappy employee stays, the more entrenched they become, and so it would take a bigger offer to dislodge them.

"We want people working at Amazon who want to be here," Amazon spokesperson Melanie Etches told CNBC Make It. "In the long-term, staying somewhere you don't want to be isn't healthy for our employees or for the company."

Of course, Amazon being Amazon, there's more to their offer than meets the eye.

Many analysts believe Amazon benefits from The Offer more than you might think. While it removes employees who are unsatisfied with their position and who may not be giving their all to the company, it also saves the company money in the long run.

The company has long made it clear that fulfillment center jobs should be high turnover. They *want* you to quit after a year or two. The churn allows the company to replace older workers with fresh new hires at starting wage with no benefits. It saves the company money in stock payouts, 401(k) payouts, vacation pay, regular and overtime pay, virtual card payments, and other benefits.

This means that while Amazon incurs a relatively small expense when paying for workers to quit, in the end the savings to the company more than make up for the initial loss.

Here's the bottom line: If you truly want a happy worker, then when you bring him or her into your organization, don't just say, "Sink or swim." That's not productive. Instead, treat them like family. Show them some love and prove that you want them to succeed. Model respect and good teamwork.

4. The Mid-Career Touchpoint

You need employees who are loyal, happy, and want to stay with your company.

To be more specific, you want a robust complement of two distinct types of employees:

1. **The Leader**, who wants to climb the corporate ladder and eventually assume a position of leadership, either as CEO or on the executive team. The quintessential examples are CEOs like Mary Barra (GM), Doug McMillon (Walmart), and Chris Rondeau (Planet Fitness). These and many more very happy leaders started at the very bottom of their companies and worked their way up to the top.

2. **The Homemaker**, who does not aspire to leadership but instead wants to rise to a certain level of responsibility and stay there. There are many good people who are happy workers who don't want the responsibility of running the company or planning for an uncertain future. Such people can be invaluable in terms of "organizational memory" and providing continuity, but they may also be resistant to change. They may have outside interests, such as church, family, or a hobby, and look at their job as a way of providing for their family and having a respectable community identity.

Of course, you need the other types as well, but they're likely to be more transient.

To keep mid-career employees happy, you need to provide a few things.

Loyalty. They're ready to extend loyalty when they feel they're getting it in return.

Respect. While many mid-career Homemaker types aren't interested in shouldering increasing responsibilities, they absolutely want their service to be respected. If they have an opinion, they want it to be heard and acknowledged.

A happy workplace. If your job is a stepping stone to something else, you'll endure more hatepoints because you know you're just passing through. But if you're going to spend a good chunk of your life at one company, you need to feel happy about coming to work in the morning. And it's not about free bagels in the break room; remember the formula:

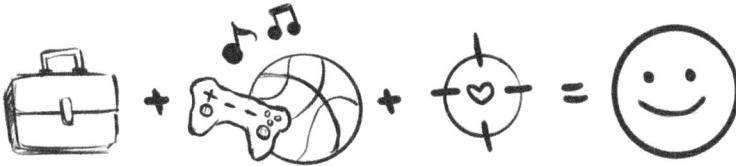

Work + Recreation + Meaning = Happy Work

When you have the right mix, you'll have an employee who's not thinking about jumping ship.

Competitive pay. Anyone who chooses to work for one company at the same level for years knows they're not going to get rich. You can safely assume that such workers are not driven solely by the quest for more money. They have other things that are important to them. Respect this, and don't take it for granted. Pay—including profit sharing—that keeps pace with the competition is very important.

Lifelong learning. Every company experiences chaotic change while (hopefully) fostering innovation. Each and every employee needs to be provided with the appropriate training and education to maintain the flexibility and strength of the company through periods of change. This helps the organization perform better and staves off the boredom that inevitably comes from doing the same thing for long periods of time.

For Leaders who seek to climb the executive ladder, you need to provide an upward route. Obviously, this will have limitations. Every organization resembles a pyramid, with increasingly fewer opportunities as you ascend higher. Not every Leader can find a role for themselves! You need to have clear plans of succession, so that when (for example) a vice president leaves for another firm that offers a leadership role, you've got a manager ready to take the promotion and fill the role that will be vacant.

Mentorship can be an important asset for grooming future leaders. In a program of leadership development coaching, a good mentor can help a subordinate or younger executive come out of his or her comfort zone by creating different leadership scenarios.

Mentors can provide specific insights and information that enable the mentee's success. They can also help in networking and introduce the mentee to the people and organizations they'll need to interact with as they climb the organizational ladder.

Even better, the benefits of mentoring go both ways. While leadership mentoring is primarily for the benefit of the mentees, by giving back to the people in their organization, leaders who choose to mentor others derive personal fulfillment through their contributions. They may also find it rewarding to see the employees they mentor succeed in their careers.

5. Post-Retirement

Many leaders foolishly think, "After Smith or Jones retires, what do I care about them? They're gone. They're ancient history."

This is very short-sighted thinking. It pays to stay on good terms with them, because you never know when you might need them.

During the waning of the Covid-19 pandemic as workplaces tried to ramp up to full speed, suddenly employers experienced a severe labor shortage. This crisis—the term is not used lightly—was caused by a variety of conditions that, like a perfect storm, all came together at once: worker disillusionment with their former jobs, aversion to commuting again, fear of Covid infection, lack of affordable childcare, and many schools operating remotely. In a tight labor market, many employers sought every solution when seeking to fill open positions. Many turned to their own former employees and other retirees. According to Rand Corporation, of workers who were age 65 or older, 40 percent had retired at least once before.

According to a study conducted by job search platform Resume Builder, 20 percent of retirees said past employers asked them to return because of the labor shortage. And while 41 percent said they would consider going back to their former position, 59 percent indicated they wanted to seek employment elsewhere.

Of those who didn't want to return to their former employer, a majority said they wanted to switch industries, with the most common reasons being less stress, the ability to work remotely, and switching to a career they were more interested in.

Some retirees returned to work because they needed money to pay their bills, while others used their paycheck for travel, hobbies, or leisure activities. Many wanted to work because they found it fulfilling and a more appealing alternative to other retirement options.

By bringing back a retired employee, companies can save time on training and compensation. Retirees returning to the company reduces the need to recruit for an employee who would have to learn the company's processes. It also allows employers to pay retirees a reduced rate, since it's often unnecessary for them to be employed full time and they're probably taking Social Security.

Oh, and what about the idea that older employees are resistant to learning new techniques and approaches? As Alison Pearson, head of HR at a personal injury law firm in Pittsburgh told SHRM, "There's a stereotype that the older an employee, the less likely they are able to adapt to technology and grow with the company. But in my experience, that's a lazy perspective."

Older workers may also have years of experience a variety of career fields, which can make them more versatile than younger workers who are just starting out in their chosen field. A highly experienced older worker can serve as a mentor and be an invaluable source of knowledge to support for younger workers.

Your former employees are a valuable "reserve force" that can help you when you need them. It's all the more reason to make them happy in the years *before* they retire!

Take Action!

✗ The employee's journey with your company comprises a series of five stations or touchpoints. It's critically important that you understand it and ensure that at every step your employee is happy and therefore productive.

✗ The sum total of the employee experience over the five touchpoints—which may extend in time to cover many years—is the Net Employee Experience (NEX).

✗ The five touchpoints are:

1. **The Pre-Touch**, which is the awareness of your business in the mind of the future employee.

2. **The Interviews.** You need to design an interview process that will give the right candidate a very good feeling about the organization and its direction.

3. **Onboarding and Training.** How you handle the first few days and months of a new employee's experience is crucial to ensuring high retention. Onboarding new hires should be a strategic process that lasts at least one year. You may even want to consider a "pay to quit" offer.

4. **The Mid-Career Touchpoint.** To keep these employees happy, you need to provide loyalty, respect, a happy workplace, competitive pay, lifelong learning. For Leaders who seek to climb the executive ladder, you need to provide an upward route.

5. **Post-Retirement.** It pays to stay on good terms with your retired employees, because you never know when you might need them. Your former employees are a valuable "reserve force" that can help you when you need them.

Thank You!

Here's more information that can help you leverage *Chaotic Change* for spectacular success!

We're in This Together

I'm always honored when a reader reaches out asking for additional insights or recommendations. The work that I do is a privilege, and I am always happy to help my reader community get to a better place. Please reach out if I can be of any service to you at nick@goleaderlogic.com.

Free Chaotic Change Mastermind Course

This powerful program is delivered in an audio format, allowing you to work out or commute while taking the Mastermind Training Program. For information on how to access this short hour and 15 minute mastermind audiobook, simply visit the book resource page at www.nickwebb.com/books.

CHAOTIC CHANGE

MASTERCLASS AUDIOBOOK

NICHOLAS J. WEBB

Resources for Winning in a Time of Chaotic Change

Understanding the DNA of chaotic change is critical for leaders to guide their enterprise to sustainable success and future readiness. However, you may need additional insights, resources, and training to optimize your strategy moving forward. With that in mind, I've listed the following important resources to ensure your success.

Get Better Insights and Strategies

If your organization needs assistance in improving the insights necessary to drive innovation in a time of chaotic change, please visit our website for resources and service options at www.goleaderlogic.com.

Events and Meetings

You can book me as a speaker for your next event, meeting, or executive leadership program.

Simply reach out through our website for information about having me speak at one of your upcoming events at www.nickwebb.com.

Chaotic Change Workforce Development

We offer over a dozen certification training programs to help our clients lead their market. These programs cover strategic planning, innovation, customer experience, professional communications, quality of work life, and leadership, to name just a few. For more information, visit www.mylearnlogic.com.

Other Services Leader Logic Provides

Innovation Management, Fractional Chief Innovation Officer, Innovation Scouting, Strategic, Planning, Executive Leadership Retreats, Customer Experience (CX) Consulting, Human Experience (HX) Consulting, Leadership Development, Keynote Speaking, Executive and Leadership Coaching, Product Launch Planning, full suite of services for Membership-Based Organizations. www.goleaderlogic.com.

It is a remarkable privilege to have time-constrained leaders and executives dedicate their moments to engage with my research and content. I extend my deepest gratitude to you for reading this book and entrusting me to provide insights that I hope will have a positive impact on your role as a leader and on your organization.

With sincerest thanks,

Nicholas J. Webb

Notes

1. "The Hands-On CEO," QSR, https://www.qsrmagazine.com/growth/denise
   -lee-yohn-qsrs-marketing-guru/hands-ceo/.
2. CNBC, "5 Key Business Lessons from Amazon's Jeff Bezos," https://www
   .cnbc.com/2016/05/13/5-key-business-lessons-from-amazons-jeff-bezos.html.
3. Marguerite Ward, "Former Google Career Coach Shares 3 Ways to Ace Your
   First Week at a New Job," CNBC, February 22, 2018, https://www.cnbc
   .com/2017/12/29/former-google-career-coach-shares-ways-to-ace-your-first
   -week-at-work.html.
4. Jeanine Prime and Elizabeth Salib, "The Best Leaders Are Humble Leaders,"
   Harvard Business Review, May 12, 2014, https://hbr.org/2014/05/the-best
   -leaders-are-humble-leaders.
5. Ira Kalb, "Innovation Isn't Just About Brainstorming New Ideas," Business
   Insider, July 8, 2013, https://www.businessinsider.com/innovate-or-die-a
   -mantra-for-every-business-2013-7.
6. "Who the Hell Wants to Hear Actors Talk?" Quote Investigator, November
   29, 2016, https://quoteinvestigator.com/2016/11/29/actors-talk/.
7. Robert Strohmeyer, "The 7 Worst Tech Predictions of All Time," *PC World*,
   December 31, 2008, https://www.pcworld.com/article/155984/worst_tech_
   predictions.html.
8. "Great Moments in Business Forecasting," http://leeds-faculty.colorado.edu/
   moyes/bplan/forecast.htm.
9. Bill Murphy Jr., "7 Things to Remember When You Lose Confidence,
   Courtesy of Airbnb," *Inc.*, July 14, 2015, https://www.inc.com/bill-murphy
   -jr/7-people-turned-down-the-chance-to-invest-in-airbnb.html; Brian
   Chesky; "7 Rejections," Medium, July 12, 2015, https://medium.com/
   @bchesky/7-rejections-7d894cbaa084.

10. CCL. https://www.ccl.org/

11. "Leadership in the Age of AI," Infosys.com, https://www.infosys.com/age-of-ai.html.

12. Rob Wile "A Venture Capital Firm Just Named an Algorithm to Its Board of Directors—Here's What It Actually Does," *Business Insider*, May 13, 2014, https://www.businessinsider.com/vital-named-to-board-2014-5.

13. Nicky Burridge, "Artificial Intelligence Gets a Seat in the Boardroom," Nikkei Asia, May 10, 2017, https://asia.nikkei.com/Business/Artificial-intelligence-gets-a-seat-in-the-boardroom.

14. Sophi Brown "Could Computers Take over the Boardroom?" CNN, October 1, 2014, https://www.cnn.com/2014/09/30/business/computers-ceo-boardroom-robot-boss/.

15. Hans Moravec, *Mind Children*, Boston: Harvard University Press, 1988.

16. Teramind, https://www.teramind.co/.

17. Veriato, https://www.veriato.com/products/veriato-cerebral-insider-threat-detection-software.

18. ActivTrak, "Productivity Reports," https://www.activtrak.com/product/productivity-reports/.

19. Praharsha Arnand, "Microsoft Patents Tech to Combat Employee Stress," ITPro.com, April 28, 2021, https://www.itpro.co.uk/business-strategy/careers-training/359370/microsoft-patents-tech-to-combat-employee-stress.

20. Jodi Kantor, "The Rise of the Work Productivity Score," *New York Times*, August 14, 2022, https://www.nytimes.com/interactive/2022/08/14/business/worker-productivity-tracking.html.

21. Neil Hodge, "Big Business or Big Brother? The Risks of Employee Monitoring," November 2, 2020, http://www.rmmagazine.com/articles/article/2020/11/02/-Big-Business-or-Big-Brother-The-Risks-of-Employee-Monitoring-.

22. Ibid.

23. Thomas W. Malone, Robert Laubacher, and Chrysanthos Dellarocas, "The Collective Intelligence Genome," MIT Sloan Review, April 1, 2010, https://sloanreview.mit.edu/article/the-collective-intelligence-genome/.

24. Thomas W. Malone, Robert Laubacher and Chrysanthos Dellarocas, "Harnessing Crowds: Mapping the Genome of Collective Intelligence," MIT Sloan School, April 16, 2009, https://papers.ssrn.com/sol3/papers.cfm?abstract_id=1381502.

25. "SHRM Reports Toxic Workplace Cultures Cost Billions," SHRM, September 25, 2019, https://www.shrm.org/about-shrm/press-room/press-releases/pages/shrm-reports-toxic-workplace-cultures-cost-billions.aspx

#:~:text=NEW%20YORK%20%E2%80%94%20One%20in%20five ,SHRM%20report%20on%20workplace%20culture.

26. Kevin Stiroh, "The Economics of Why Companies Don't Fix Their Toxic Cultures," *Harvard Business Review*, March 22, 2018, https://hbr.org/ 2018/03/the-economics-of-why-companies-dont-fix-their-toxic-cultures.

27. Nick Otto, "Employee Stress Costing Employers Billions in Lost Productivity," March 25, 2019, EBN, https://www.benefitnews.com/news/ employee-stress-lost-productivity-costing-employers-billions.

28. "12 Mind-Blowing Employee Survey Statistics," Officevibe, September 9, 2014, https://officevibe.com/blog/employee-surveys-infographic.

29. Nishal Mistri, "Stop Wasting Money on Benefits That Employees Don't Value," LinkedIn, April 10, 2018, https://www.linkedin.com/pulse/stop -wasting-money-benefits-employees-dont-value-nishal-mistri/.

30. Teresa M. Amabile and Steven J. Kramer, "The Power of Small Wins," *Harvard Business Review*, May 2011, https://hbr.org/2011/05/the-power-of -small-wins.

31. Axelle Tessandier, "The New Explorers," *Huffington Post*, January 23, 2014, https://www.huffingtonpost.com/axelle-tessandier/the-new-explorers _b_4275782.html.

32. Polly Mosendz, "Microsoft's CEO Sent a 3,187-Word Memo and We Read It So You Don't Have To," *Atlantic*, July 10, 2014, https://www.theatlantic .com/technology/archive/2014/07/microsofts-ceo-sent-a-3187-word-memo -and-we-read-it-so-you-dont-have-to/374230/.

33. HBR. https://hbr.org/1989/09/speed-simplicity-self-confidence-an-interview -with-jack-welch.

34. "How to Write Vision and Mission Statements for B2B. And Why It Matters," https://www.themarketingblender.com/vision-mission -statements/; "100+ of the World's Best Vision Statements," https://www .executestrategy.net/wp-content/uploads/2016/06/100-of-the-worlds-best -vision-statements.pdf.

35. "The Innovation Gap," Korn Ferry Institute, https://www.kornferry.com/ content/dam/kornferry/docs/article-migration/Korn-Ferry-Institute-the -innovation-gap.pdf

36. "Initial Franchising Costs with 7-Eleven," November 21, 2018, http:// franchise.7-eleven.com/franchise-blog/initial-franchising-costs-with-7 -eleven; https://www.entrepreneur.com/franchises/7eleveninc/282052.

37. Andrea Gabor, "Seeing Your Company as a System," *Strategy+Business*, May 25, 2010, https://www.strategy-business.com/article/10210?gko=20cca.

38. Max Blau, "No Accident: Inside GM's Deadly Ignition Switch Scandal," Atlanta, January 6, 2016, https://www.atlantamagazine.com/great-reads/no-accident-inside-gms-deadly-ignition-switch-scandal/; Wikipedia "General Motors Ignition Switch Recalls," https://en.wikipedia.org/wiki/General_Motors_ignition_switch_recalls.

39. "ANA Quality and Innovation Conference—Sharing Innovations and Advice," American Nurses Association, March 23, 2018, https://www.nursingworld.org/news/news-releases/2018/ana-quality-and-innovation-conference---sharing-innovations-and-advice/.

40. Ibid.

41. "Steve Jobs Speaks Out," CNN Money, March 7, 2008, https://money.cnn.com/galleries/2008/fortune/0803/gallery.jobsqna.fortune/3.html#:~:text=%22We%20do%20no%20market%20research,never%20hire%20consultants%2C%20per%20se.

42. Susan Hanley, "As Steve Jobs Once Said, 'It's Really Hard to Design Products by Focus Groups . . .,'" Computerworld, April 7, 2008, https://www.computerworld.com/article/2343679/as-steve-jobs-once-said---it-s-really-hard-to-design-products-by-focus-groups--.html.

43. Marine Aubagna, "How Customer Surveys Help Apple Maintains Industry Leadership," Skeepers, August 10, 2020, https://skeepers.io/en/blog/how-customer-feedback-surveys-helps-apple-maintains-industry-leadership/.

44. Ryan Gould, "Survey Fatigue: What It Is and How to Avoid It," SurveyAnyPlace.com, February 3, 2021, https://surveyanyplace.com/survey-fatigue/.

45. Brooke Sjoberg, "TikTok Shows Superior Telling Amazon Workers Not to Pack up Early—1 Minute Before the End of Their Shift," DailyDot, August 4, 2021, https://www.dailydot.com/debug/amazon-workers-told-not-to-pack-up-early-tiktok/.

46. Elsi Boskamp, "25+ Crucial Average Cost Per Hire Facts [2023]: All Cost of Hiring Statistics," Zippia.com, February 16, 2023, https://www.zippia.com/advice/cost-of-hiring-statistics-average-cost-per-hire/.

47. "Why Zappos.com Pays New Hires $1,000 to Quit" Amanet, March 24, 2020, https://www.amanet.org/articles/why-zappos-com-pays-new-hires-1-000-to-quit/.

48. Ruth Umoh, "Why Amazon Pays Employees $5,000 to Quit" CNBC, May 21, 2018, https://www.cnbc.com/2018/05/21/why-amazon-pays-employees-5000-to-quit.html.

49. Kylie Lobell, "Welcoming Back Retired Employees" SHRM, August 5, 2021, https://www.shrm.org/resourcesandtools/hr-topics/employee -relations/pages/retirees-return-to-work-.aspx.

50. "Labor Shortage Driving Demand for Retirees—20% Have Been Asked to Return to Work," ResumeBuilder, March 22, 2024, https://www. resumebuilder.com/labor-shortages-driving-demand-for-retirees/.

51. Kylie Lobell, "Welcoming Back Retired Employees," SHRM, August 5, 2021, https://www.shrm.org/resourcesandtools/hr-topics/employee -relations/pages/retirees-return-to-work-.aspx.

About the Author

Nick Webb is a number one bestselling author and one of the top keynote speakers in the world. Nick speaks on the future of innovation, healthcare, technology, innovation, leadership, and the changing workforce. As the CEO of LeaderLogic, LLC, Nick serves some of the top brands in the world, helping them to build innovative and future-ready enterprises.

Nick began his career as a technologist, inventing one of the world's smallest medical implants for the treatment of ocular surface disease. Nick has been awarded over 40 patents for technologies that include one of the first wearable technologies, and a wide range of industrial and consumer products. Nick's bestselling books include *What Customers Crave, What Customers Hate, The Innovation Mandate, The Healthcare Mandate, Lucid Leadership*, and *Happy Work*, just to name a few. Nick has also served as an Adjunct Professor at a medical school where he also served as the Chief Innovation Officer. Nick also leads LearnLogic, a training firm that provides certification and custom training to organizations around the world.

Nick is also an award-winning documentary filmmaker. His film *The Healthcare Cure* recently won the People's Choice Award at the prestigious International Sedona Film Festival.

In his work as a speaker, Nick travels the world, sharing his latest research to help organizations build future-ready enterprises in a time of rapid change. Nick is also extremely passionate about human-centric organizations that create cultures of happiness for both employees and the customers they serve. Nick also operates a small lab where he works on technologies specific to continuous patient monitoring, 3-D printing, and healthcare applications for artificial intelligence.

Made in the USA
Monee, IL
23 August 2025